Bulk water pipelines

Tim Burstall

Thomas Telford

Published by Thomas Telford Publications, Thomas Telford Services Ltd, 1 Heron Quay, London E14 4JD

First published 1997

Distributors for Thomas Telford books are
USA: American Society of Civil Engineers, Publications Sales Department, 345 East 47th Street, New York, NY 10017-2398
Japan: Maruzen Co. Ltd, Book Department, 3–10 Nihonbashi 2-chome, Chuo-ku, Tokyo 103
Australia: DA Books and Journals, 648 Whitehorse Road, Mitcham 3132, Victoria

1156296 X

Learning Resources
Centre

A catalogue record for this book is available from the British Library.

Classification
Availability: Unrestricted
Content: Original analysis
Status: Author's invited opinion
User: Civil and pipeline design engineers

ISBN: 0 7277 2609 9

© Tim Burstall, 1997

Typeset in Great Britain by Alden Bookset, Oxford.
Printed in Great Britain by Galliard (Printers) Ltd, Great Yarmouth, Norfolk.

Preface

The purpose of this book is to provide some critical perspectives on facets of bulk water pipelines. You, the intended readers, are pipeline engineers, involved in operations and maintenance, pipeline design engineers, and field supervisors and personnel involved in construction, operations and maintenance. I hope you will incorporate those perspectives and accompanying ideas that are valid for you in your design, construction, operation and maintenance processes. The book is not intended to be used as a design manual as each waterworks organisation will most likely have its own manual.

The material comes largely from my work as Pipeline Engineer in the Bulk Water Department of the Wellington Regional Council, New Zealand (hereafter RCW). I was fortunate to be the first and only holder of this position. Previously, I had worked as a Construction Engineer, with a strong accent on pipeline systems in the oil, petro-chemical, and pulp and paper industries. A fourth generation Mechanical Engineer, I entered the world of water, dominated by civil engineers, with more than a little trepidation. I have endeavoured to bring some fresh perspectives to this work that I love. The accent is practical and the approach is intuitive.

Thanks are due to the Wellington Regional Council, New Zealand, for providing the many photographs and drawings. Thanks also to the pipeline equipment manufacturers who have provided illustrations and comments on the text. Thank you to those who have answered my many questions over the years; in particular, the field staff of the RCW Bulk Water Department who have implemented, modified and improved some of my ideas, as well as my engineer colleagues in the RCW who have listened and provided critical feedback. Thanks also to R.A. (Bob) James of Cathodic Protection New Zealand Ltd,

Wellington, for the provision of the Cathodic Protection Specification in Chapter 6, and Derek Neale for the property perspectives of Chapter 15. Thanks also to Craig Ramsay of Deeco Ltd, New Zealand, for the introduction on control valves in chapter 5. If there are perspectives I have overlooked, missed, or misrepresented, I welcome your comments.

Tim Burstall
Wellington, New Zealand

Contents

Introduction

Bulk water pipelines are the connecting network between headworks, i.e. intakes and treatment plants, and the reservoirs which each supply a specific area of a town or city. The reticulation pipes which connect to reservoir outlets are *not* directly considered in this book. These larger outlet mains are similar in materials and construction to bulk water mains so that some of the perspectives may be directly applicable.

The presence of earthquakes in New Zealand and elsewhere reminds us of the many *lifelines* that are keeping our cities healthy. Bulk water mains are very important lifelines providing a primary supply of treated water to the cities they serve. Each reservoir has a limited storage volume. This equates to a given time, varying with demand, in which to maintain or modify the inlet line(s) without interrupting the supply to consumers. When the water system was built, this time could well be 1·5 to 2 days, perhaps more. In today's setting, with the evolution of each system catering for the growth in demand, the storage time might be less than 1 day. This fact requires a slick, well-organised pipeline section to keep the customers satisfied.

Such an operation is essential if the bulk water pipelines are subject to earthquakes. Much more is required and I urge you to read the section on Earthquake Response Planning in the Appendix, which provides specific perspectives on this subject.

I include below a humorous specification for pipe purchase, which some readers may not have read!

Pipes: purchase and work specifications
1. All pipe is to be made of a long hole surrounded by metal or plastic.

2. All pipe is to be hollow throughout the entire length. Do not use holes of a different length than the pipe.

3. The I.D. (inside diameter) of the pipe must not exceed the O.D. (outside diameter), otherwise the hole will be on the outside.

4. It is desirable that the hole in the pipe be the same diameter for the entire length, otherwise the contents of the pipe will get thinner in the smaller part of the pipe.

5. All pipe is to be supplied with nothing in the hole so that water, steam or other stuff can be put in at a later date if required.

6. All pipe should be supplied without rust. This can be readily applied at the job site. Note: Some suppliers will supply pre-rusted pipe. This should be cheaper as it is usually old stock.

7. All pipe over 500 ft (153 m) in length should have the words 'long pipe' clearly painted on each end, so that the contractor will know it is a long pipe.

8. Pipe over 2 miles (3·2 km) in length must have the words 'longer pipe' painted in the middle, so the contractor will not have to waste time walking the entire length to determine whether it is a long pipe or not.

9. All pipe over 6 in (150 mm) in diameter must have the words 'large pipe' painted on it so that the contractor will not mistake it for a small pipe.

10. Pipes that are threaded on both ends should have either right hand or left hand threads. Do not mix the threads, otherwise the coupling is screwed on to one pipe as it is unscrewed from the other pipe. This could cause the pipe to be disconnected and then leak.

11. Pipes that are to have flanges must have them on both ends. Flanges must have holes for the bolts that are quite separate from the big hole in the middle.

12. When joining two pipes together with flanges, it is necessary to have flanges of the same size, otherwise the bolts will not fit into the holes and the pipes will not stay together, causing a likely leak.

13. When ordering 90 degree, 45 degree or 30 degree elbows, be sure to specify right hand or left hand — otherwise you will end up going the wrong way and the required 180 degree elbows are an added cost.

14. When ordering pipe, be sure to specify to your supplier whether you want level, uphill or downhill pipe. If you use downhill pipe for going uphill, the contents will flow the wrong way.

15. Check if the pipe includes ready made leaks. If not, these may be added during the construction process. Most pipes leak from the

inside out but do not be surprised if the occasional one leaks from the outside in. In some instances where leaks are not desirable, to check if the pipe is working, it may be possible to use both types of leaks together, to offset each other.

Pipe dimensional parameters

Pipe is specified by two size parameters, namely diameter and wall thickness, and two physical parameters, namely material specification and method of manufacture. Pressure rating of pipe is sometimes specified but this is a derived parameter, calculated from the diameter (D), wall thickness (t), and allowable stress (S), by the basic hoop stress equation

$$S = p(D/2t)$$

The description of pipe size by nominal diameter can be confusing, especially in the smaller sizes. Further confusion occurs because imperial sizes are referred to using a metric description, derived from an imperial conversion, then rounded to a whole integer. For example, $\frac{1}{2}$ in pipe is called 15 mm. A 20 in diameter pipe is referred to as 500 mm pipe, though is actually 508 mm outside diameter, a direct conversion of inches to millimetres being $20 \times 25 \cdot 4$. The pipe, then, is imperial in metric clothing.

The most perplexing description of pipe diameter is nominal diameter (nominal bore, NB), which is the way pipe is described and specified. Nominal diameter is often mistaken to mean inside diameter, and for relatively thin wall pipes, this assumption is approximately correct. For example, 6 in (150 mm) nominal schedule 40 pipe has a bore of 154·08 mm, very close to 6 in (150 mm).

However, 6 in (150 mm) schedule XXS has a bore of 124·5 mm, very close to 5 in (125 mm). Schedule is a system of specifying particular standard wall thickness. These dimensions are derived from ANSI B36.10, ANSI B36.19 or BS 1600. How can a 6 in (150 mm) pipe have a 125 mm bore yet still be called a 6 in (150 mm) pipe?

Nominal (diameter) is derived from the Latin *nomen* which means name. Nominal diameter is therefore **only** a convenient label or name tag. The outside diameter (D_0) of a pipe is the **actual** size parameter. For standard pipe, the outside diameter always corresponds to the nominal diameter. For example, 6 in (150 mm) nominal pipe is always 168 mm outside diameter specified under ANSI B36.10. This is true regardless of the pipe wall thickness. The inside diameter or bore (D_i) of any pipe, having a wall thickness t, is a calculated, derived parameter according to the equation

$$D_i = D_0 - 2t$$

D_i is of interest because it determines flow area, and a design engineer will set a minimum area (and hence inside diameter). For a lined pipe with a lining thickness t_l, the equation below governs the inside diameter

$$D_i = D_0 - 2(t + t_l)$$

Here, the inside diameter is (smaller) because the pipe is lined, yet the actual pipe, lined or not, is identical. This further illustrates the folly of associating nominal diameter with the **same** inside diameter. Confusion can always be avoided by also always specifying the outside diameter. This is the **critical parameter** because couplings must be specified to suit. The fitting clearance on a coupling is small so the coupling must match the pipe outside diameter. Likewise, a weld band (socket type welded coupling) must match the outside diameter of the pipes to be joined. Specifying outside diameter always avoids the confusion caused by multiple historical and present pipe standards. Tables detailing these standards and the different pipe outside diameters are available and very useful.

Pipe has been and can be made to non-standard (outside) diameters, e.g. rolled steel 972 mm. This is justified economically on long new pipelines which can be made exactly to the size determined by the designer. This practice does, however, defeat the purpose of having standards and creates the practical problem of providing discrete (special) sized repair materials in stock. Another approach for historical non-standard pipes is to standardise replacement pipes and to employ transitional couplings (mechanical or welded). For example, 933 mm outside diameter pipe is in the ground; 914 mm standard pipe and welded thickening rings 933/914 can be used along with 933 mm couplings left over from the original construction. Alternatively, 933/914 step couplings can be used.

Some plastic pipes, e.g. uPVC, are manufactured to the old imperial standard diameters used for cast iron, but also are made to a true metric standard. The advantage of a true metric standard is that the nominal diameter for most sizes is the same as the outside diameter, e.g. 200 mm, not a conversion from imperial. Small plastic pipe, e.g. MDPE, is typically 20, 25, 40, 50 and 63 mm outside diameter. The 63 mm outside diameter is 2 in (50 mm) nominal diameter following BS 1600. However, it is better to refer to it as 63 mm pipe to avoid confusion with 50 mm outside diameter (40 mm NB). For couplings on plastic pipe using inserts, the wall thickness as well as the outside diameter must be specified to enable the insert to fit (inside) the pipe.

1. Pipeline materials

The Pipeline Engineer and field staff must be familiar with the range of pipe materials that are both used and *able to be used* on bulk water duties. In addition, historical materials still in use, e.g. cast iron, must be understood. Finally, we all have to get to know the technological field of plastic pipelines, which is the area of growth. Plastics are encroaching on applications which previously were served by metallic materials. The growth and development of polyethylene (PE) pipe is a prime example.

The selection of new materials is a difficult subject, fortunately made easier by P.J. de Rosa's book, *Pipeline materials selection manual*, published by the WRC.[1] This gives a method which yields various options depending on the parameters inputted into the algorithm. The pipe material options, e.g. steel, ductile iron, prestressed concrete and asbestos cement (AC), cannot be simply equated. Cost comes into the equation but other preferences may prevail. There are no hard categories of right and wrong.[2] Each material has its strengths and weaknesses and the author has tried to show these qualities and give some specific applications to suit their strengths.

1.1. Plastics

Plastic pipe is well accepted for water reticulation. Several different materials are used. All types are corrosion resistant, ductile and relatively impervious. However, some people are concerned that external contaminants may be able to enter the water.[3] The strength of plastic materials is much less than steel or ductile iron and this limits the size available at high pressures. Thicker walls in plastic pipe are used to compensate for this lower strength. Some asset managers and engineers still think plastic is on 'trial'. Typical asset life is 50 years,

Fig. 1. Hobas GRP pipe (courtesy James Hardie Pipelines Ltd, New Zealand)

though this is probably a conservative estimate by the manufacturers. Cities in the Wellington region in New Zealand are installing plastic reticulation mains operating at 90 m pressure. Elsewhere,[4] glassfibre reinforced polyester (GRP) bulk mains have been installed (Fig. 1). Other uses for plastic pipe are raw water lines in catchment areas and raw water lines on artesian wellfields. Spans between supports on above ground plastic pipes are shorter than steel, so extra supports will be required. For replacement pipe crossings in rugged, difficult to access, land, plastic may not be the best choice. Stainless steel, schedule 10s, is a good alternative for this application, say up to 250 mm diameter.

Plastic pipes are suitable for scour valve (drain valve) tail pipes. Finally, plastics are most suited for chemical lines at treatment plants. For example, conveying corrosive sodium hypochlorite solution into bulk mains for disinfection purposes.

More care is required when constructing a plastic pipeline as it is more easily damaged, e.g. by careless backfilling. One advantage in smaller diameters is longer lengths and therefore fewer joints. Plastic pipes can easily be installed inside older, larger cast iron mains, instead of replacement. The author does raise the possibility of failure in a seismic event, resulting from cast iron shards penetrating the plastic.

Plastic U-liners are another choice. Solvent welded (e.g. uPVC) or fusion welded (e.g. PE) plastic pipes and fittings form an integral system. Rubber ring jointed systems on plastic pipes (e.g. GRP, uPVC) are analogous to ductile iron systems. Plastic pipes can also be jointed using mechanical couplings (refer 3.2.2). Note, PE pipes must use a coupling providing axial restraint.

Polyethylene pipe can be joined by fusion welding using two methods, namely butt welding and electrofusion couplings. The butt welding process is analogous to metal welding (refer 3.1.1). The electrofusion method uses resistance wires built into the special coupling. A timed electrical current heats the coupling and pipe to give a controlled temperature and consistent weld.

1.2. Ductile iron

Ductile iron pipe, concrete lined, is the natural successor to cast iron pipe (Fig. 2). It is a ductile material more like steel than iron but with good resistance to external corrosion. Engineers are still assessing the properties of ductile iron pipe which is only about 30 years old. It is claimed in accelerated soil corrosion tests that the life expectancy is similar to cast iron. The concerns of engineers about the adequacy of the factory-supplied polythene bag and whether or not to apply cathodic protection (CP), show that ductile iron is still on 'trial'. A ductile iron pipeline in the USA was tested for CP currents before and after bonding across the joints. The authors conclude that protective currents are more than an order of magnitude higher than for well-coated (tape) steel pipe.[5] A recent installation in the USA specified two polyethylene bags and a CP system, because of high chloride content in the surrounding soil.[6] Another different approach by a UK Water Authority was to target only the ductile iron pipe joints with a specially developed coating product.[7]

Ductile iron fittings are available to form a complete system. However, owing to the unique angles of most site bends, prefabricated steel bends (specials) are often used with mechanical couplings. A short branch line might end up 25% steel, 75% ductile iron. This is not a unified system. However, an advantage of the steel bends is that they can be fitted with leak detection bars to facilitate future leak detection. Long straights in ductile iron will need the occasional steel section to allow welding of such bars.

Standard ductile iron pipes are provided with spigot and socket ends. There are special joint types which provide full axial restraint, thus avoiding the need for anchor blocks on bends etc. Repair of a leaking rubber ring joint can be achieved by cutting out a section and splicing in a new section (e.g. steel) with two mechanical joints — a relatively expensive procedure. Branches require a similar procedure, i.e. a steel special. It is possible to weld it successfully, but ductile iron welders will likely be hard to find.

Conceptually, a ductile iron pipeline system is different from a fully welded steel pipeline. Any joint can potentially leak, whereas a modern double lap or butt weld gives a strength equal to the parent pipe. Thus a welded system can be considered a unitary pipeline, effectively having no joints.

At the time of writing, imported ductile iron pipe was cost competitive up to 375 mm diameter. In the larger sizes, the material costs are similar although steel construction costs are higher owing to the welding required.

Fig. 2. Ductile iron pipe (courtesy Tubemakers of Australia)

1.3. Asbestos cement

Asbestos cement (AC) was invented around 1930 by Dr Mazza, an Austrian. Constructed of several layers of asbestos fibres soaked in cement, it was thought to be impervious to corrosion — especially electrolytic corrosion: 'AC pipes are reasonably resistant to corrosion, although being a cement-based product, they are susceptible to attack by acidic, soft or sulphate-bearing conveyed water and soils. The use of bitumen dipped pipe is recommended for Water Industry applications'.[8] Pipe ends are plain, and fit sockets (collars) with rubber ring seals. AC pipelines use cast or ductile iron bends and fittings; the cast iron can sometimes be unlined. AC is a brittle material and fails catastrophically. Sometimes, an excavator is not required for repairs as the burst scours its own hole, often large enough for access.

Asbestos cement is still used in the USA and has recently passed the scrutiny of USEPA. The asbestos fibres in the water are not known to cause any health problems. However, AC is linked to the toxicity of asbestos as far as public perception is concerned. In some countries, it is an historic material, superseded by steel, ductile iron, and plastics. AC pipe walls swell over time, indicating there can be an internal reaction. Asset life is around the 50 year mark.

Pipeline location of an AC main is difficult. Historically, AC was constructed without a metallic tracer and it cannot therefore be located by a pipeline detector. Practitioners will have to use as-built plans and hand dig extra carefully to locate the pipe. This is a hit and miss method which can result in a great number of small location holes in a roadway. A cheaper and more accurate method is to cut the AC pipeline at regular intervals and to send a video camera with a sonde up the pipe. The location of the main can then be traced on the surface.

'Location may also be accomplished using buried pipe detection apparatus designed to locate non-metallic buried pipelines (such as asbestos-cement and plastic). Such devices operate in a manner similar to a simple seismograph and have proven to be effective. Probing with metal bars is not recommended'.[9] This method also requires connections to the pipe in order to transmit the pressure pulse, e.g. hydrants. AC bulk mains would have to be retrofitted with such connections at suitable intervals, a costly, but nevertheless feasible, project. Another technology for locating non-metallic pipes is ground penetrating radar, which provides a cross-section of the ground traversed.

Repair of burst AC mains is usually achieved by splicing in a new section of pipe (e.g. steel). Two mechanical couplings are needed.

Differences in diameter between the AC and the replacement pipe can be catered for by using a variable fit coupling (up to 300 mm dia.) or step couplings purchased specifically for each application. Small splits and leaks can be repaired with mechanical repair clamps.

1.4. Steel — mild

Steel used in water pipelines is inherently strong, yet ductile. It is easily worked and welded. Corrosion is the main problem, causing pinhole leaks and most likely loss of wall thickness in the pipe. Cathodic protection preventing external corrosion should be an integral part of steel water pipeline design. It should not be an add-on as it can be expensive to retrofit (refer to Chapter 6) The weldments on steel pipes 40 years or older will likely be of poorer quality than present-day welds. However, failure of welds requires abnormal stresses induced, for example, by earthquakes.

Steel has a ductile/brittle transition zone just below 0 °C. This means its ability to absorb energy is low and its ability to blunt cracks and prevent them from growing is reduced. Older steel water lines in cold climates can potentially fail catastrophically at the welds *under normal operating conditions.*[10]

Modern mild steel offers the engineer the biggest choice of repair methods. Pinhole leaks can usually be plugged with wooden stakes, hammered in under pressure. The wet, expanded stake is then cut off close to the main. Finally, a stainless steel threaded socket is centralised over the leak hole and welded to the pipe wall, and the socket plugged off.

Large splits in steel pipes can be repaired if curved plates are welded over the affected area, or a repair clamp can be fitted. The old methods of rubber wedged against the leak under a girth clamp should not be used for permanent repairs as inevitably the leak eventually reappears. A welded method is the best as it retains the integrity of the pipe wall.

Rubber ring joint (RRJ) pipe with a failure at the joint is repaired easily by cutting out the joint and installing a replacement section of pipe with two mechanical couplings.

1.4.1. Spiral riveted pipe

Spiral riveted pipe dates from the turn of the century. Welding was in its infancy and presumably not considered adequate for pressure pipe. Mechanical means were therefore used. Hemp was used in the joints of some older pipe. The mechanical lap joints are strong and do not generally leak. Corrosion generally occurs away from the joints and can be welded, or a repair clamp can be used. A leak on a riveted joint might be fixed with a repair clamp or, in the worst case, a full replacement length of pipe will have to be installed between two existing mechanical couplings.

1.4.2. Lockbar pipe

Lockbar pipe dates approximately from 1920 to 1940. Ends are plain for the fitting of gibault type couplings or the occasional flange. To make this pipe, steel plates were rolled, then their ends upset to form a shoulder.[11] The two plate ends were crimped together and fitted into either side of an I-bar steel section, the full length of the pipe (Fig. 3). The outside 150 mm of I-bar, adjacent to the pipe ends was cut off flush with the pipe so formed. The plates were welded to the cut down I-bar to create rounded ends. Then, the mechanical seal between the plates and the I-bar was made by squeezing the top of the I-bar against a supported, internal anvil.

Lockbar pipe is strong. One pipeline encountered by the author had no record of leaking at the mechanical seal over a 70 year period (31 km of pipe). Modifications to the pipe, e.g. installation of branches, can prove difficult. There is a maximum size of branch that can be

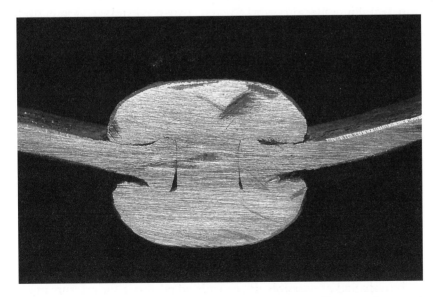

Fig. 3. Section through lockbar pipe (courtesy Wellington Regional Council, New Zealand)

fitted practically, as under normal pressures a compensating ring (plate) will be needed. A further problem may be the orientation of the lockbar (random) interfering with the line of the proposed branch. Then, the whole length has to be dug up and rotated to allow the branch to be fitted. If this is the case, it is better to replace the length with modern pipe.

The author and a colleague developed a method to cut lockbar pipe and to reinstate the round end, in order to facilitate concrete relining without having to remove a full length at each access pit. The process of trimming the outside of the I-bar back and welding the plates to the bar causes the lockbar to expand and leak under pressure. Welding the lockbar first also causes leaks. So a liquid compound was applied to the outside of the lockbar. This reduced the leakage to a few small weeps on the test pipe. Finally, a cement mortar lining was hand applied and tested at twice the working pressure, with no leakage evident. One further problem occurred when the lockbar pipe was cut and reinstated. The cut ends were not round but resembled a polygon. The pipe had been made presumably with a flanging process rather than with a rolling process. Only the 150 mm long plain ends were round. The problem was then sealing the pipe ends with a circular mechanical joint. Sleeve couplings with larger rubber rings worked, with some juggling of the bolt tightening sequence. Lip seal couplings also worked well, two being tested to just over three times the rated working pressure before they began slowly to weep.

These techniques allow the Pipeline Engineer to modify lockbar pipes without full lengths between couplings having to be replaced.

1.4.3. Rolled pipe

Rolled pipe is made by rolling plate into a cylinder and joining the seams. Riveted seam rolled pipe was first used in the USA in the early 1850s.[12] Some early welded pipes were lap welded, not butt welded. Their plate thickness is higher than that of a butt welded longitudinal seam to allow for a welded joint efficiency of 0·8 or so. Spigot and socket ends for welding provide approximately 2 degrees of deflection before welding. This avoids the use of bends (specials) for small changes in angles.

Smaller diameters, i.e. less than 750 mm, will likely be plain ended for mechanical couplings. Plain ends may also be used for larger diameter pipes with mechanical couplings (e.g. pipe through a tunnel). Repairs to rolled pipe are easily made by welding.

1.4.4. Spiral welded pipe

The natural successor to spiral riveted pipe, spiral welded pipe, is fully fusion butt welded along the spiral seam (Fig. 4). Ends can be furnished plain, spigot and socket for welding (slip joint) or spigot and socket rubber ring (RRJ). Diameters 600 mm and above can be welded from the inside as well as the outside (e.g. double lap) and the internal lining repaired to retain the integrity of the lining. Cutting a full length to fit a mechanical coupling requires grinding the proud spiral seam flush with the outside of the pipe. Otherwise, spiral welded pipe can be treated as identical to rolled pipe.

1.4.5. Seamless pipe

Seamless pipe is made from a solid billet of steel, which is pierced to form a hollow round section and then drawn through dies to finished

Fig. 4. Spiral welded steel pipe machine (courtesy Humes Steelpipe, New Zealand)

dimensions. Seamless pipe is inherently strong as it has no welds. It is more expensive than fabricated pipe and is used often in the petrochemical industries.

1.4.6. Galvanised pipe

Small diameter steel pipes are often referred to as galvanised iron but are in fact galvanised steel. British Standard 1387 for medium galvanised pipe was used extensively for small diameter water pipes on city reticulation service pipes. The complementary screwed fittings are made of galvanised iron. Some of these pipes may exist on your bulk water pipeline system, perhaps on small bypasses or pressure gauge lines. The life expectancy of this pipe is anywhere between 25 and 50 years. The interior blocks over time through tuberculation, and the outside of the pipe can rust very quickly in corrosive soils (buried situation).

The weakness of threaded systems is at the threaded joints. A buried galvanised pipe will have to be dug up to be repaired. It is not practical to field thread in situ, in order to make a repair. So a mechanical coupling must be used. Pipes above ground can be repaired by uncoupling the nearest union, or fitting an extra union to allow pipes to be reinstated.

On bulk mains, the operating pressures may be higher than the rating of the galvanised iron fittings (typically 140 m max). If this is the case, heavy wrought steel screwed fittings could be used. However, other materials are superior.

1.4.7. Grooved end lightweight galvanised pipe

This hot dipped galvanised pipe is available up to 250 mm diameter or so. Ends are grooved to fit two-piece couplings (refer 3.2.2.2.3). The life expectancy is 50 years based on above-ground use. In corrosive soils, without a tape wrap, the life would be less. A possible application in a bulk supply system is *temporary lines*, particularly after an earthquake. It is lighter than concrete-lined steel and therefore much easier to install. The grooved end system takes full axial load across the joints so that anchor blocks are not required, an essential feature of temporary works.

1.5. Stainless steel

Electric fusion welding (EFW) stainless steel conforms to ANSI dimensions and it therefore differs in diameter from water pipes 500 mm diameter or less. Adaptor couplings will therefore be required to connect in stainless steel. Available in several grades, the ones most likely to be encountered are 304, 316, and 321. The L grade (low carbon content) is the normal grade supplied, to allow welding. The advantages of stainless steel are its corrosion resistance and light weight. No lining or coating is required. The cost of imported 150 mm dia. schedule 10s 316 EFW was identical to locally made 150 mm spiral welded steel pipe, at the time of writing. The author specified 316 grade stainless steel pipe for replacement of 100 m of 150 mm dia. steel pipe in the pavement duct of a bridge across a sea inlet, carrying a major highway. Push-in RRJ pipe would have fouled the underside of the pavement slabs (40 mm clearance). Lip seal mechanical couplings were used which cleared the bottom of the duct (15 mm clearance) and the underside of the pavement slabs. Two people comfortably carried a 6 m length of stainless steel pipe.

In more normal situations, stainless steel is more costly as it requires welding or mechanical joints. The advantage mild steel has is the easy provision of a RRJ in the pipe at the factory, which costs little more than plain ended pipe.

Stainless steel is a good candidate for sections of the bulk water pipeline system that are difficult to fabricate or to replace and for applications requiring durability;[13] for example, stream crossings. The material is virtually both zero maintenance and vandal proof. Unusual applications for stainless steel pipe are raw water pipelines in catchment areas. These lines should be flushed with water periodically to prevent sediment from building up and causing undersediment corrosion. Stainless steel should be carefully considered for stagnant sections and pipes not flowing fully.[13] The author knows of several 200 mm diameter pipelines in densely forested areas accessed by walking tracks. These pipes date from 1924 and are of spiral riveted steel construction, bitumen lined and coated. A recent slip has destroyed 25 m of pipe installed against the stream bank. Replacement in stainless steel would be a good option. It is able to span the existing supports and is not too heavy to install manually. Low pH waters are no problem, and a long asset life can be expected. Should repair of the stainless steel be necessary, welding is generally the first option and a mechanical solution the second option.

1.5.1. Stainless steel — spiral wound

Spiral wound stainless steel pipe can be made locally to almost any diameter and is likely to be more readily available than EFW imported pipe.

1.5.2. Stainless steel — small diameter

Small diameter concrete lined steel pipes (15–50 mm) do not exist, for obvious reasons. The natural candidate for these lines is stainless steel schedule 40, which is the minimum thickness to allow threading of the pipe. Connections to bulk mains can be made by welding a stainless steel threaded socket or half coupling to the main pipe. Small diameter threaded pipework can be fitted to form a small bypass or to connect to a single air valve. Stainless steel is superior to galvanised steel (corrosion) and copper (mechanical strength) in these applications. In New York and Tokyo, stainless steel is being installed for service pipes.[13]

Threaded pipe systems, as mentioned previously, can leak at the joints. Why not eliminate all threaded joints between the bulk main and all threaded branch valves? This procedure is used in the petrochemical industries for obvious maintenance and operation reasons. In one instance experienced by the author, 4 km of 1 m diameter water main had to be shut to remake a 50 mm leaking threaded joint. Such a shutdown may not be possible on other bulk water pipeline systems.

Flanged valves can be used instead of screwed valves but they are more expensive and remove only one threaded joint on the small diameter inlet pipe. Instead, stainless steel butt weld fittings, pipes and flanged valves can be used. These fittings can be tungsten inert gas (TIG) or manual metal arc (MMA) butt welded in the workshop or field. This will result in a fully integral corrosion-resistant small diameter pipeline up to the first branch valve.

1.6. Concrete

Prestressed concrete pipe is used in the USA at high pressures (up to 160 m approx.). Conceptually, it is similar to a ductile iron system, which is its natural competitor. The comments on ductile iron thus apply. An issue concerning the use of prestressed concrete pipe is corrosion of the steel component. Cathodic protection can mitigate against this. Concrete pipe is heavy and usually comes in short lengths. This means there are more joints to make than other systems.

Leaks in the joints of concrete pipes can be repaired with lip seals fitted internally across the leaking joint, and, in addition, as in the case of Northridge CA 1994, an external repair clamp installed.[14] Otherwise, the standard method of pipe and two couplings can be used.

Precast concrete pipe similar to stormwater pipe but made to a higher quality can be used for low pressure (up to 15 m) duties, such as raw water intake lines. Connections are made by pushing the spigot end into the adjacent socket end as in ductile iron RRJ pipes. Non-standard bends and fittings can be fabricated from steel concrete-lined pipe and connected to the concrete pipes with mechanical couplings.

Concrete does not normally corrode from the outside. Raw waters can corrode the interior, exposing the reinforcing steel. Further corrosion might cause structural failure, creating large leaks. Repairs to the interior can be made with epoxy compounds, but the cost may be prohibitive. Alternatively, plastic pipe or U-liner can be installed inside the concrete.

1.7. Cast iron

The earliest authentic recorded use of cast iron water pipes dates back to 1455, the pipe being of German origin.[15] Cast iron pipes installed in 1664 still grace the Palace Gardens at Versailles, France. Generally, cast iron pipes exhibit good corrosion resistance. On the inside, rust tubercules form and require periodic scraping to remove them. In-situ cement mortar lining, or other methods of lining cast iron pipes, can be an economic solution to extend the asset life. This also improves the flow and water quality for approximately one third to one quarter of the cost of a new main. Alternatively, chemical cleaning may be possible (refer 2.1).

Before relining is undertaken, it is advisable to check the extent of the internal and external corrosion in order to determine the 'good' wall thickness. Stress analysis will show if the good metal left is sufficient. If not, sections of the cast iron will need to be replaced. A full survey of such a main may not be possible for operational reasons. If so, sample sections of the main can be removed for further investigation.

Cast iron is a brittle material and fails catastrophically. If the leakage history shows frequent breaks, it is worth considering a replacement of the whole line. Leakage at lead-packed joints is another operational problem (refer 3.8).

Cast iron pipes cannot practically be welded. Small diameter connections are made with tapping bands. Large diameter, e.g. 200 mm, connections can be made by installing a short steel section in the main with a 200 mm dia. steel branch. This requires two mechanical couplings. Alternatively, a stainless steel clamp with built-in tee might be used (refer 3.2.2.3.2). Ovality in the cast iron can test the abilities of the couplings to seal (refer 3.2.2.1.3).

Failure of cast iron pipes can be attributed to rapid pressure variations, e.g. surges. Culprits might be control valves which open/close too quickly (refer 5.1.1) and malfunctioning air valves. The operation of manually operated line valves should not cause problems if carried out under the proper waterworks procedure.

Circumferential cracks in cast iron flanges at the pipe/flange fillet cannot be repaired. A replacement steel fabrication is recommended. Small diameter failed cast iron fittings can be refabricated in stainless steel, which avoids the need for linings and coatings.

Longitudinal cracks run between spigot and socket joints. In this instance, both joints will have to be cut out and a length of pipe (e.g. steel) spliced in using two mechanical couplings. The cast iron can be cut by machine cutters which are rolled around the pipe circumference on tracks, wrapped round the pipe. Adequate clearance around the pipe is needed for machine cutters and this is not always available. Other methods to cut cast iron can be used such as pad saws or unguided power saws with circular cutting discs, although the latter method is not as safe.

1.8. Copper

Copper has a long association with potable water, dating back to 2300 BC or earlier. It is widely used for internal water pipes in domestic and commercial buildings, although plastic pipes are making inroads into these areas. Copper can corrode internally under certain conditions and copper alloy fittings can dezinc, thus causing failure. This is not common.

In a bulk water system, copper can be used for small impulse lines, e.g. orifice tapping lines, pressure gauge lines, etc. For above-ground use, e.g. in a pump station, copper is a good material. In buried service, it is more susceptible to physical damage and external corrosion. Copper may have been used to connect 25 mm air valves to bulk mains. The author traced the cause of a mechanical failure in copper pipe to the large weight of a particular model of cast iron air valves. Presumably, the failure occurred through movement of the whole assembly, even though buried. Stainless steel schedule 40 provides much more mechanical strength and is a superior material for this application.

2. Linings and coatings

2.1. Linings

Linings are applied to the interior of pipes to provide corrosion resistance and a hygienic barrier between the potable water and the pipe wall. All linings should be periodically inspected, with particular emphasis on the joints (refer chapter 7). Linings are made of the following materials:

bitumen
coal tar
coal tar epoxy
epoxy resin
cement mortar
paint systems
polyethylene.

Chemical cleaning of unlined pipes having tuberculation, is a possible alternative to relining (as in 2.1.6, 2.1.8).[16] The authors state that the same problems are being found in uPVC, PE and AC pipes. The economics of this process may not be viable for long, large diameter bulk mains because of the large volume of chemical required. Units capable of cleaning 304 m of 200 mm diameter pipe are available. The valves used to isolate the pipe section to be cleaned must be tight. The process is quick compared to scraping and relining.

2.1.1. Bitumen

Bitumen linings are made up of approximately 80% of bitumen and 20% of dry lime. Bitumen deteriorates with age. One 70 year old pipeline, managed by the author, showed some loss of lining and a puffing-up of the bitumen to give a rough tuberculed appearance. Raw water lines of similar age had lost most of their lining.

One problem with bitumen-lined pipes removed from service and stored above ground is cracking, which can cause chunks to fall off. This appears to be caused by evaporation of water absorbed during service. Repair of the lining is necessary before reuse. This can be achieved in a large diameter pipe through the application of hot bitumen to the cleaned, dried and primed surface. Repairs in situ, on pipes in the ground, can be tackled likewise. If the area to be repaired is a long way from an access point, a threaded socket can be welded to the pipe locally to allow oxy-acetylene hoses and cables for power tools to pass through. After repair, the socket is plugged.

Welded joints with cracks and loose lining in the original bitumen repair can be fitted with lip seals as an alternative to in-situ bitumen repairs. The lip seal here is not sealing a leak but is sealing the faulty lining in the joint, from the water.

2.1.2. Coal tar enamel

Coal tar enamel (CTE) was applied to pipes in the 1960s and earlier, but is no longer used on health grounds (leaching of aromatic carcinogens) and water quality grounds (supports microbial growth affecting taste and quality).[17] Cement mortar linings are now favoured. One advantage of CTE is its resistance to low pH waters. Accordingly, CTE was used on raw water lines such as artesian wellfields. Field joints on large welded pipes were made by inserting a mould inside the pipe and pouring hot CTE through a threaded socket previously welded on the pipe joint. The socket was plugged afterwards. These joints constitute a potential weakness in the lining and therefore periodic inspection is advisable.

Coal tar enamel can be repaired in a large diameter pipe, much as bitumen (refer above). CTE can be trowelled on to a primed, clean, dry surface by heating with a blowtorch. Extraction of the fumes would be necessary by opening several access points and using forced ventilation. A respirator must also be worn. Present Health and Safety laws may prohibit such repairs. If so, repairs might be possible by applying epoxy compounds or sophisticated compatible paint systems.

2.1.3. Coal tar epoxy

A coal tar epoxy lining system has been used on self-supporting stream crossings to reduce overall loadings. It is known to leach undesirable chemicals into potable water and is not a preferred material.

2.1.4. Cement mortar

Cement mortar is the standard lining on ductile iron, steel and relineable cast iron. However, readers should note other linings, e.g. resins or plastic inserts, can be used to reline steel or cast iron. Cement mortar-lined pipe on raw water duty should be periodically checked to see if the thickness has reduced or if the cement mortar is corroding.

Cement mortar lining on treated water duty promises a long life. Correct chlorine residuals result in a clean surface with only a little fine dust found on internal inspection. Tests on cement mortar-lined pipe have been carried out with holes deliberately cored out, and these tests have proved that the lining bridges the hole and prevents leakage. Thus cement mortar-lined pipe should not pinhole in service, *unless* the lining has been damaged, e.g. by an excavator bucket. Excavators can do serious damage to pipe, and linings can easily be broken away from the pipe wall. This naturally leads to unseen internal corrosion and ultimately to a leak which appears to be an external pinhole leak.

2.1.4.1. Repair of cement mortar

Cement mortar linings are easily repaired using modern epoxy mortar compounds or cement mortar with additives. The epoxy compounds generally need 24 hours to cure, whereas the additive mix cures within an hour. The additive mix is the first choice for pipes that need to be back in service quickly. For workshop fabrications, the epoxy compounds are more suited to repair the lining at welds. Flanges are often necessary on pipework purely to facilitate repair of linings. The length of the person's arm carrying out the lining repair governs the position of flanges. Flanges are needed on branches added to a main pipe, for the same reason. The length of the branch up to the flange needs to be within an arm's length.

Cement mortar linings can exhibit problems. One problem encountered by the author was the formation of hard white lumps, some as deep as the lining thickness, caused by faulty plasticiser in the mix. These lumps have to be chipped out and the lining repaired.

Repairs to linings at original pipeline welds are well worth checking when the pipe interior is accessible, as these construction repairs can easily have been forgotten.

2.1.5. Paints

These are likely to be used on raw water pipelines, e.g. artesian lines, either directly on the steel or on top of a cement mortar lining. The compatibility of the paint to potable water is the key issue. It is possible for paint to flake off a cement mortar-lined wellfield pipeline. The author experienced one such pipeline which was charged but not flowing. This resulted in a foul smelling chemical entering the water and causing a taste problem (hydrocarbons). The remedial work consisted of scraping and relining the existing cement mortar lining with a further layer to allow for expected corrosion by the low pH water.

Repair of painted linings should be possible in situ, if the diameter is large enough and sufficient ventilation is achievable. Derusting and drying the affected surface will be the hardest task, and several days' curing time may be required for older paint systems. Modifications to pipelines having painted interior linings obviously require repair to the lining. Relining of the pipeline (refer 2.1.6, 2.1.8) if it is a short length may be more practical, in view of the restraints described above.

It is possible that epoxy resins approved for potable water could be used as a thin paint film.

2.1.6. Epoxy resin
The asset life of epoxy resin linings is quoted at 50 to 100 years by the manufacturers. The advantages of resin lining are:

(a) tight quality control of the finished lining
(b) fast curing time
(c) no early taint problems which are often associated with cement linings.

Resin linings are not usually offered by manufacturers on new pipes as they are not competitive with cement mortar linings. However, resin linings are used to reline watermains, especially reticulation mains where cement mortar linings would build up high alkalinity in the water and lead possibly to plumbosolvency. Fosroc, however, has developed a new, centrifugally controlled, applicator which allows pipes of up to 625 mm to be relined using epoxy resin (Figs 5 and 6).[18] Cement mortar relining is cheaper above 200 mm diameter or so, and has a long, known asset life.

2.1.7. Sintered polyethylene
This is a factory application where heated (steel) pipes are dipped in a bath containing polyethylene material (in a powder form). Thus the finished pipe is both lined and coated with one 'skin' (Fig. 7). These

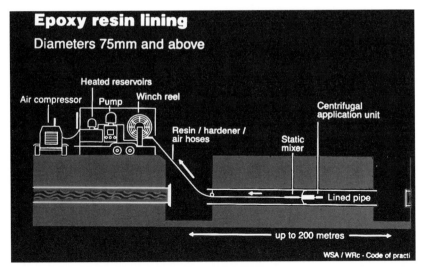

Epoxy resin lining

Diameters 75mm and above

Heated reservoirs

Air compressor Pump Winch reel

Centrifugal
application unit

Resin / hardener /
air hoses

Static
mixer

Lined pipe

up to 200 metres

WSA / WRc - Code of practi

Fig. 5. Resin relining schematic (courtesy Fosroc Ltd, New Zealand)

pipes are especially suited for sewers which have aggressive contents.
They can also be used for raw water duties. The lining process can
be applied separately from the coating process. A sintered polyethylene
pipe can therefore have a different external finish, e.g. paint.

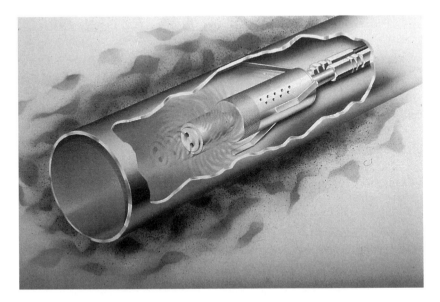

*Fig. 6. Exploded view of resin relining machine (courtesy Fosroc Ltd,
New Zealand)*

2.1.8. Relining of pipelines

Relining of old water pipelines is a growing industry on account of the cost of replacing old with new. In dense urban areas, the cost of the 'hole' has been estimated at 80% of the total pipeline construction cost. It makes sense to use this space in the ground with minimal disruption to the streets above.

There is another in-situ method, for smaller diameter cast iron pipes, which uses the space in the ground provided by the pipe, namely pipe cracking. A mole shatters the cast iron pipe into small pieces and also pulls a new pipe (e.g. plastic) into position behind the head of the mole.

The relining of old bulk water pipelines requires access pits approximately every 200 m. On larger diameter lines, use can be made of existing access points although it is unlikely cast iron pipes will have these features. Branch connections will have to be deconstructed to repair the main/branch lining interface if the main pipe is not large

Fig. 7. Sintered polyethylene lining and coating machine (courtesy Tubemakers, Australia)

enough to allow internal access. After relining has been completed, the sections of pipe removed for equipment access are replaced, usually with new cement mortar-lined steel pipe and mechanical couplings. These short sections can be fitted with leak detection bars (refer 8.3) to facilitate leak detection for the pressure test, and for future operation. Following the pressure test, the relined pipeline is disinfected using the same procedure as for a new pipeline.

2.2. Coatings

In order to evaluate the performance of pipeline coatings, the specific properties that relate to field performance must be evaluated. The criteria for pipeline evaluation are:

- resistance to permeation
- adhesion
- impact resistance
- indentation resistance
- soil stress resistance
- ductility
- resistance to deterioration
- cathodic protection current requirements.

The information given in Table 1 has been extracted from AS 2832.1 — 1985: *Guide to the cathodic protection of metals, Part 1 — Pipes, cables and ducts.*

There is value in the periodic inspection of coatings undertaken by digging up sections of pipeline and inspecting the coating. This procedure is mandatory on petrochemical pipelines. Pipelines protected cathodically can be surveyed, and coating defects indicated can be dug up, inspected and repaired. The defect size and location can be correlated with the instrument readings as a cross check. Cathodic protection measurements should be interpreted to see if the coating is deteriorating. The periodic dig up, however, is still recommended as a reality test on the coating.

2.2.1. Bitumen, coal tar enamel

The comments on bitumen and coal tar enamel (CTE) linings apply equally to coatings.

2.2.2. Gunite

The gunite coating consists of concrete sprayed on to the pipe, which has been wrapped with steel mesh to allow build up of the concrete. The pipe

Table 1. Typical properties of coating systems for buried structures

Coating system	Coating site	Structure pretreatment (steel)	Coating thickness, mm	Susceptibility to damage from		
				Soil stresses	Cathodic disbondment	Impact
Coal tar enamel	Over ditch or shop	Wire brush* or blast	2·5 to 5·0	Medium	Medium	Medium
Extruded polyethylene	Shop	Blast	0·6 to 1·8	Low	Low	Medium
Sintered polyethylene	Shop	Blast	1·6 to 2·3†	Low	Low	Low
Coal tar epoxy	Site and shop	Blast	0·3 to 0·5	Low	Medium	Low
Fusion bonded epoxy‡	Shop	Blast	0·2 to 0·3	Low	Medium	Low
Asphaltic enamel	Over ditch and shop	Wire brush* or blast	2·5 to 5·0	Medium	Medium	Medium
Butyl laminated tapes	Over ditch	Wire brush*	0·7 to 3·0	High	High	High
Petrolatum tapes	Over ditch	Wire brush*	1·4 to 5·0	High	High	High
Heat shrink sleeve	Over ditch	Blast	1·0 to 3·0	Low	Medium	Medium

* Wire brush pre-treatment which leaves millscale on the surface may leave steel in a condition susceptible to stress-corrosion cracking and is inferior to blast cleaned surfaces for good adhesion of the coating.
† Fusion bonded medium density polyethylene coating and lining for pipes and fittings (AS 4321, 1995).
‡ Fusion bonded epoxy resins are usually applied within a thickness range of 200–600 µm, depending on service requirements.

weight is likely to be heavier than pipe with other coatings. Pipes covered in concrete should suffer virtually zero corrosion.

2.2.3. Petrolatum tape system

The petrolatum system is likely to be used on mechanical couplings and valves, rather than the full pipeline. It can be used on above-ground or below-ground pipelines. The 'Denso' system consists of a petroleum-based paste applied as a primer to the metal to be protected. Mastic is applied to improve the contour of the profile for the smooth application of tape. A final protection layer of tape can be applied to seal the inner wrap. The petrolatum system has a major advantage in its flexibility, allowing application to difficult profiles. The mastic layer helps in this regard as noted above. The system can also be applied to damp surfaces (refer 2.2.7).

The petrolatum system is well regarded by engineers and the author has seen bright, bare steel, 30 years old, which had been protected and showed no signs of corrosion.

2.2.4. Blue jacket

The blue jacket coating is factory applied on cement mortar-lined steel pipe, typically up to 300 mm nominal diameter. It consists of a bitumen primer layer, and an extruded polyethylene coating — hence the name blue jacket, blue being the signifier for water pipes. It should be noted that this coating has been supplied previously in black jacket. There are questions about the porosity of the coating, as cutting it back for a fabrication sometimes reveals sweating on the steel pipe surface and minor rust spots.

2.2.5. Tape systems

Tape systems consist of three layers: primer; inner wrap; outerwrap. The primer/inner wrap adhesion interface constitutes the primary corrosion barrier. The outer wrap provides some mechanical protection. The tape materials are made from polyethylene and butyl rubber. The two different material layers are laminated together. One company which calls its product a coating system uses an extrusion method to combine the two different material layers to give one tape. The two layers cannot be prised apart. Tape systems are shop applied after pipe manufacture. In addition, tape wrapping can be done in the field and repairs to tape systems can be carried out.

Damage to tape systems on pipes can occur easily during transit or loading/offloading. Sharp objects can easily cut through all the layers.

Soft slinging of tape-wrapped pipe is essential and quality backfill without any large rocks is needed in the trench.

The outer tape can be subject to degradation by ultraviolet light. This is a problem for spare pipes kept for repairs. Longer term ultraviolet resistance for the outerwrap can be specified, but additional black polyethylene sheeting wrapped around the pipe is recommended. Alternatively, the pipes can be stored under cover. This tends to be expensive. Another solution for all spare pipes in the immediate repair stock is to cover them with bedding material, having first capped the ends.

A development of the tape system is heat treatment of the tape-wrapped pipe, which is supposed to give a very hard surface, virtually immune to damage.

2.2.5.1. Above-ground tape systems
A long-life reduced-maintenance alternative to painting is a reinforced tape wrap system overcoated with epoxy resin.

2.2.6. Sintered polyethylene
Sintered polyethylene is applied in the factory by dipping heated (steel) pipes in a bath containing the polyethylene coating material (in a powder form). The resulting pipe coating is very hard and durable. Sintered polyethylene coated pipes resist ultraviolet light, and pipe stored in the open does not require any protection. On rubber ring jointed pipes, it is applied around the ends so that the pipe is fully protected on the outside and needs no further wrap at the joint. Field repairs to the coating can be made on site. A full range of factory fabricated and coated fittings is available so that every element of the pipeline system has a shop-applied sintered polyethylene coating.

2.2.7. Paint systems
Paint systems are usually used on exposed pipelines, such as stream crossings (Fig. 8). For pipes near the sea, a good marine paint system should be chosen, typically inorganic zinc primer and a two-part epoxy undercoat and gloss. Regular inspection of the pipes, say every two years, is required. This is because marine environments can attack the welds on pipes and the pipe wall itself rather quickly. Colours, if chosen well, can make the pipe as unobtrusive as a large pipe can be. Remedial work is best left for the summer, when the steel temperature is highest; this is because, in winter, the pipes can sweat continuously. For pipes under bridges, it may be necessary to

Fig. 8. Stream crossing (courtesy Wellington Regional Council, New Zealand)

drain them to prevent sweating on the steel. If this is not possible, a petrolatum tape system can be applied. An outer protection may be needed, in addition, to protect against vandalism.

Paint systems are unfortunately easily vandalised. A tape system with outer mechanical protection may be the answer, or the system of 2.2.5.1 above. Alternatively, stainless steel pipe can be used for small diameter pipe crossings.

3. Joints

Joints in a pipeline system are unavoidable because the supplied lengths of pipe are finite, typically between 6 m and 12 m for metallic pipe. One large diameter (450 mm and 17 bar rating) plastic pipe was supplied recently in a 90 m length, with flanged ends.[19] This method significantly reduces the number of joints.

Joints fall into three categories: welded; mechanically coupled joints; and flanged joints which are a mixture of welding and mechanical means. The pressure rating of joints must be greater than the maximum operating pressure of the pipeline under consideration, at the point of use. This is important when modifying existing lines by adding flanges, now supplied under a modern metric standard, e.g. BS 4504.

Joints are a potential source of leakage and accordingly the number of them should be minimised.

3.1. Welded joints
Welded joints comprise the following types of joint

- butt joints
- slip joints
- double lap joints — weld bands.

The main advantage of welded joints is that the completed pipeline is essentially *one piece*. 'Thrust blocks are not required on steel pipelines when sufficient anchoring force due to pipe friction is available.'[20] However, thrust blocks may be fitted at the construction stage to anticipate later modifications or repairs using mechanical couplings. Welded joints are always accompanied by joints in the lining. Linings can be repaired on pipes large enough in diameter to allow internal access (refer 2.1.4.1) Smaller diameter pipe joints will require epoxy mortar applied to the inside of the joint before being assembled for welding and a swab pulled through after welding.

3.1.1. Welding processes

Various welding processes can be employed to weld steel pipes together, whether in the workshop or field. Manual metal arc (MMA) is the typical process used but metal inert gas (MIG), tungsten inert gas (TIG) or oxy-acetylene welding can also be used. BS 2971 Class II arc welding of carbon steel pipework for carrying fluids is an appropriate standard to use for bulk water pipelines operating up to 24 bar pressure. The companion standard for gas welding is BS 2640.

The first step in any welding process is to establish an approved welding procedure, e.g. BS EN 288-2. The purpose of the welding procedure is to prove the finished material properties of the particular weldment produced. A welding engineer will formulate the welding procedure which should contain the following:

- drawing of the weld undertaken, showing pipe diameter and thickness
- fit up details
- welding process to be used
- electrode type, number and size of weld runs
- typical amperage for the welding runs
- any requirements for pre-heat or post-heat treatment of the weld.

The welding procedure is proved by various tests, non-destructive and destructive (mechanical). The standard used, e.g. BS EN 288-3, for the welding procedure will define the acceptable material properties of the weld. The proving of a welding procedure also automatically qualifies the person who produced the weld.

It should be noted that mild steel welding procedures are standard procedures in the construction industry and the test is usually waived by principals.

The second step is welder qualification. This is a similar test, e.g. BS EN 287-1, to the welding procedure but the pipes (or plates) to be joined are in various spatial positions; the 45° horizontal and vertical positions (6 g) being the most difficult. There is a hierarchy of welding positions such that the qualification at any level is automatic qualification at any lower levels.

Every welder working on pipeline welds needs to be qualified. Principals often accept welder qualifications from previous similar work and waive the physical weld and testing. The basis for the material in sections 3.1.1.1 to 3.1.1.6 is the course notes from the Welding Institute, Cambridge: *Weld defects and their origins*, 1972.

3.1.1.1. Manual metal arc (shielded metal arc, stick, electric arc welding)

Manual metal arc (MMA) remains the most versatile welding process and is the likely process to be used on pipeline field joints. The welding electrode contains a metal compatible with the materials being joined, with a flux which forms a shield on arc initiation to protect the weld pool from contamination. The welder adjusts the electrode feed rate by hand to keep the arc length constant. MMA produces a slag which protects the weld metal from oxidising while it is cooling. The slag must be chipped off and ground out before the next welding run (pass) is carried out. Typical power sources are three-phase AC and DC motor driven generator set (field).

The root run of the weld requires a small diameter electrode but the subsequent weld run(s) will use larger diameter electrodes with a consequent higher current (amps) setting of the welding machine.

The operating parameters are:

current range:	75–300 A
thickness range:	2 mm up
deposition rate:	1–5 kg/hr
types of joint:	all
welding position:	all (dependent on flux coating)
access:	good
portability:	good.

The consumables are:

metal rods 1·5–8 mm diameter with flux covering (1–5 mm radial thickness); the characteristics of manual-metal-arc electrodes, i.e. arc stability, depth of penetration, rate of deposition, position of welding, depend on the chemical composition of the electrode coating, classified by BS EN 499 (for mild steel electrodes).

3.1.1.2. Metal inert gas (metal arc gas shielded, semi-automatic, metal active gas)

Metal inert gas (MIG) relies on a gas shield (typically 95% argon, 5% oxygen) to protect the weld pool. The gas is provided by a gas cylinder which gently pressurises the welding torch. The welding electrode is in the form of a wire, which is automatically advanced through the centre of the torch into the weld pool. The MIG set stores the wire on a reel and has a mechanism for feeding the wire through the torch at a constant speed (current). The operator is not therefore concerned with controlling the arc length.

Metal inert gas has two modes: dip transfer; spray transfer. Dip transfer occurs at low amperage and the wire melts off in droplets to join the weld pool. Spray transfer has a high current demand and the wire melts off to join the weld pool in a spray formation. Large welds can be deposited in a single pass in spray transfer mode.

Metal inert gas on site is not common and requires protective tents to prevent loss of the gas shield from winds. MIG is a quick process giving a good weld appearance. The welding procedure has to be followed with more care than MMA to avoid faults such as lack of fusion.

The operating parameters are:

current range:	60–500 A
deposition rate:	1–10 kg/hr
thickness range:	dip transfer, pulsed arc 0·5 mm up
	spray transfer 6 mm up
types of joint:	all, including spot welds
welding positions:	dip transfer, pulsed arc — all
	spray transfer — flat only
access:	fair
portability:	fair.

The consumables are:

(a) electrodes, bare wire (mild steel is normally copper coated) 0·6–1·6 mm diameter, layer wound on spools 0·5–12 kg weight; composition of wire selected to suit parent material; wire specifications covered by BS EN 440; electrodes, flux cored are available for high deposition rate welding.

(b) shielding gas in cylinders containing compressed gas which will expand to $7 \, m^3$ at atmospheric pressure; usually carbon dioxide for mild steel, argon for non-ferrous materials.

3.1.1.3. Gas welding (usually oxy-acetylene)

The hot gases of burning fuel gas and oxygen provide the fusion heat and a protective reducing (chemically) atmosphere for the weld pool. This process heats a larger area of the parent pipe(s) than electric welding. The operator must manipulate the blowpipe to give the correct weld pool size and add bare filler wires. The process is slow compared to electric welding. The weld quality is not as good as electric welding (tensile strength of the weldment).

The operating parameters are:

gas consumption:	$0{\cdot}03{-}3\,\mathrm{m}^3/\mathrm{hr}$
deposition rate:	up to 1 kg/hr
thickness range:	$0{\cdot}5$ mm up
types of joint:	all, including spot welds
welding position:	flat or vertical
access:	good
portability:	excellent.

3.1.1.4. Tungsten inert gas (tungsten arc gas shielded, argon arc, heliarc, heliweld, gas tungsten arc welding, GTAW)

The tungsten inert gas (TIG) welding process is analogous to oxy-acetylene welding. The tungsten electrode must not be consumed and it provides the arc for the fusion weld, either autogenous (no filler rods) or with filler rods. The arc is unstable at low currents and special provision is made for starting, e.g. high frequency or surge injection. As with MIG, a protective gas shield is provided through the welding torch.

Tungsten inert gas gives a high quality weld on many materials. It is commonly used for stainless steel welding where the pipes to be joined must be purged of air to prevent the root of the weld from oxidising, and subsequently corroding. TIG is used on mild steel root welds (the critical run) with MMA fill and cap runs in the petrochemical industry. For normal water pipelines, this quality is not necessary. TIG is a slower process than either MMA or MIG, but very thin material can be successfully welded.

The operating parameters are:

current range:	10–400 A
deposition rate:	0–2 kg/hr
thickness range:	$0{\cdot}1$ mm up
type of joint:	all, including spot welds
welding positions:	all
access:	good
portability:	fair.

3.1.1.5. Pre-heat and post-heat treatment

Pre-heat and post-heat treatment are usually required when alloy steels, e.g. chrome-molybdenum, are being welded. Heavy welds in carbon steel, e.g. 20 mm up, without a low hydrogen electrode, requires pre-heat to avoid too rapid a rate of cooling of the weldment, which can cause cracking. Pre-heat is also required for the same reason when the ambient temperature is low, e.g. below 5 °C.

3.1.1.6. Weld inspection and defects

Visual weld inspection should always be the first technique used. Poor appearance of the weldment can often indicate other defects. Other non-destructive methods include crack testing (dye penetrant, magnetic, eddy current), radiographic and ultrasonic techniques. Destructive techniques which determine the mechanical strength require a sample of welding to be removed or a test coupon at the end of a weld run. The coupon may be better aligned than the weld itself and give a misleadingly high strength. The level of defects acceptable is given in the standard named in the contract documentation, e.g. BS 2971 for arc welding and BS 2640 for gas welding.

BS 499, Part 1: 1965 defines weld defects under six classifications, namely

1. Cracks
2. Cavities
3. Solid inclusions
4. Lack of fusion and penetration
5. Imperfect shape
6. Miscellaneous faults.

BS 2971, Section 7: Inspection, does not mention miscellaneous faults (arc strikes, spatter, tool marks, etc.), although arc strikes are covered in section 1.14. This probably reflects the Class II nature of the welding standard.

3.1.2. Butt joints

The butt joint, properly welded and tested, makes for a truly integral pipeline (Fig. 9). This gives the Engineer a high degree of confidence in the performance of the pipeline. Welded from the outside, internal access is nevertheless required to repair the lining. Therefore, the minimum diameter that can practically be welded and repaired is 600 mm. The wall thickness of water pipe at normal pressures (say maximum 160 m) is thinner than the schedule standard pipe used in

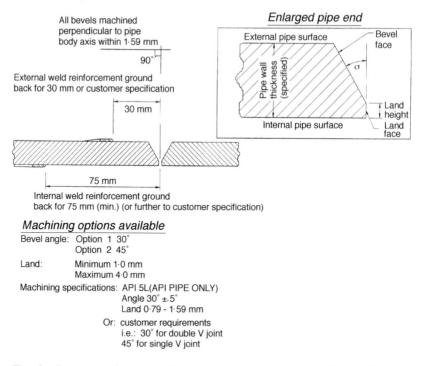

All bevels machined perpendicular to pipe body axis within 1·59 mm

90°

External weld reinforcement ground back for 30 mm or customer specification

30 mm

Enlarged pipe end

External pipe surface

Bevel face

Pipe wall thickness (specified)

σ

Internal pipe surface

Land height

Land face

75 mm

Internal weld reinforcement ground back for 75 mm (min.) (or further to customer specification)

Machining options available

Bevel angle: Option 1 30°
              Option 2 45°

Land: Minimum 1·0 mm
             Maximum 4·0 mm

Machining specifications: API 5L(API PIPE ONLY)
                          Angle 30° ±·5°
                          Land 0·79 - 1·59 mm

            Or: customer requirements
                i.e.: 30° for double V joint
                45° for single V joint

Fig. 9. Butt joint (courtesy Humes Steelpipe Ltd, New Zealand)

the petrochemical industries. Thus the out of roundness of water pipes is greater and it is more difficult to align the pipe ends for butt welding. This is why some authorities prefer the double lap joint, made with a weld band. Mitre joints for segmental bends are, of course, butt welds.

Butt welds can be inspected visually, which is always a first approach to weld testing. Radiographic (BS 2910) or ultrasonic methods can be applied to inspect the body of the weld. The non-hazardous nature of the fluid, water, determines the level of inspection required. It should be a random 10% of all butt welds.[21] This acts as a quality control on the welding. Principals can specify the level of inspection in their contract documents.

3.1.3. Slip joints

The slip joint is a circumferential fillet weld with a minimum leg length the same thickness as the pipe wall (Fig. 10). The joint is also welded on the inside if the pipe diameter allows access. Testing of the welds is carried out by pressurising the gap between the pipes, inside the socket end. Inert gas (e.g. nitrogen) is piped by way of a pressure

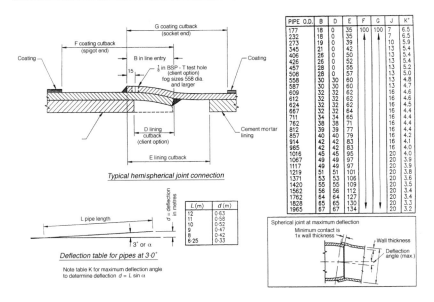

PIPE O.D.	B	D	E	F	G	J	K°
177	18	0	35	100	100	7	6.5
232	18	0	35			7	6.5
273	19	0	39			10	5.9
345	21	0	42			13	5.4
406	26	0	50			13	5.4
426	26	0	52			13	5.4
457	28	0	55			13	5.2
508	28	0	57			13	5.0
558	30	30	60			13	4.8
587	30	30	60			13	4.7
609	32	32	62			16	4.6
612	32	32	62			16	4.6
624	32	32	62			16	4.5
667	32	32	64			16	4.4
711	34	34	65			16	4.4
762	38	38	71			16	4.4
812	39	39	77			16	4.4
857	40	40	79			16	4.2
914	42	42	83			16	4.1
965	42	42	83			16	4.0
1016	45	45	95			20	4.0
1067	49	49	97			20	3.9
1117	49	49	97			20	3.9
1219	51	51	101			20	3.8
1371	53	53	106			20	3.6
1420	55	55	109			20	3.5
1562	56	56	112			20	3.4
1762	64	64	127			20	3.4
1828	65	65	130			20	3.3
1965	67	67	134			20	3.2

G coating cutback (socket end)

F coating cutback (spigot end)

B in line entry

Coating

$\frac{1}{8}$ in BSP - T test hole (client option) fog sizes 558 dia. and larger

Coating

Cement mortar lining

D lining cutback (client option)

E lining cutback

Typical hemispherical joint connection

Deflection table for pipes at 3·0°

L (m)	d (m)
12	0·63
11	0·58
10	0·52
9	0·47
8	0·42
6·25	0·33

L pipe length

d = deflection in metres

3° or α

Deflection table for pipes at 3·0°

Note table K for maximum deflection angle to determine deflection d = L sin α

Spherical joint at maximum deflection
Minimum contact is 1x wall thickness
Wall thickness
Deflection angle (max.)

Fig. 10. Slip joint (courtesy Humes Steelpipe Ltd, New Zealand)

regulator on the bottle to a small threaded hole ($\frac{1}{8}$ in) on the socket. Soapy water is applied to all the weld surfaces to test for leakage. A drop in the pressure gauge should not necessarily be interpreted as a leak because gas pressures vary with temperature.

The slip joint typically allows 2–3 degrees of deflection. Slip joints have been known to fail under compressive loads experienced during earthquakes.

3.1.4. Double lap joints

This joint is made with a welding band which is essentially a split socket that is pulled up around the joint with a bolt(s). It looks like a pipe clamp. Two external fillet welds and two internal fillet welds are made. The testing of the welds is identical to the slip joint except two test holes are required. The double lap joint is stronger than the slip joint and is preferred on pipelines subject to earthquakes. The double lap joint can be made even if the two pipes do not perfectly align, although this technique is not normally used on new construction.

3.1.5. Protection of welded joints

Welded joints can be coated with all of the coatings described in Chapter 2. The typical modern procedure is to apply a polyethylene heat shrink sleeve over the joint to provide good corrosion protection.

These are available in grades to resist soil stresses when pulling pipe, such as under river crossings inside a bored/drilled hole. Guaranteed external corrosion protection on these welds which become inaccessible, is paramount.

3.2. Mechanically coupled joints

A pipeline using mechanically coupled joints will require anchor blocks at bends, tees, etc. if the joint is unable to resist axial loads. There are several joint systems which can resist axial loads although they incur more expense. Other advantages of a mechanically jointed system are as follows.

- There is no need to repair the internal lining.
- No welding is necessary.
- There are no external coating repairs for a push-in system.
- Flexibility is built into the pipeline.

Construction costs and time are less than for welded joints. Refer 1.2 for conceptual differences.

There are two types of joint, namely

- push-in rubber ring — restrained or unrestrained
- mechanical couplings — restrained or unrestrained.

3.2.1. Push-in rubber ring joint

The push-in rubber ring joint (RRJ), often referred to as spigot and socket, is the most economical mechanical joint, as it is formed out of the pipe material at the pipe ends (Fig. 11). This makes it marginally more expensive than plain ended pipe. RRJ joints are used on metallic pipes and some plastic pipes. These joints give reliable service provided that they are properly made during pipeline construction. One problem encountered by the author was smaller diameter steel RRJ pipe of the 1960s and 1970s which had a short length socket. Some of these joints had pulled out, presumably due to ground settlement or perhaps as a result of small shakes. Younger pipes with longer engagement in the joint had not failed.

Coating of RRJ steel pipe at the joint is critical. Steel thickness in the smaller diameters is typically 3–4 mm so that poorly protected joints can corrode easily from the outside. This can occur at the groove locating the rubber seal, and failure is finally evidenced by the rubber pushing out into small holes in the pipe wall. Modern fusion bonded coated pipe avoids this problem as external protection is not

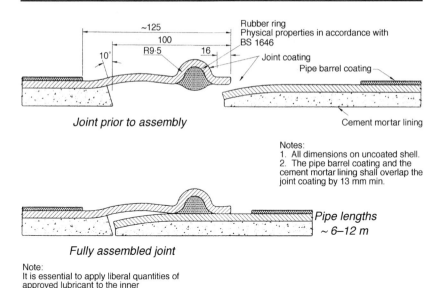

Notes:
1. All dimensions on uncoated shell.
2. The pipe barrel coating and the cement mortar lining shall overlap the joint coating by 13 mm min.

Joint prior to assembly

Fully assembled joint

Note:
It is essential to apply liberal quantities of approved lubricant to the inner surface of rubber ring and outer surface of spigot end prior to assembly. Refer to assembly procedure.

Fig. 11. Rubber ring joint (courtesy Humes Steelpipe Ltd, New Zealand)

required at the joint. Here, the coating 'wraps' around the end of the pipe to the inside lining.

The decision to use unrestrained or restrained RRJ is an engineering one. On a long straight pipeline, unrestrained RRJ is normal. For difficult terrain such as steep hills, restrained RRJ is better. Headwalls can be eliminated.

The restrained RRJ is available in ductile iron and medium density polyethylene (MDPE), but not in steel. The ductile iron RRJ restraint is a snap ring which bites into the pipes to be connected. Some designs are demountable, from the inside. 'The Japanese have developed the "S" and "SH" restrained joint that allows longitudinal movement, to a point, when the retainer ring stops pull out. When longitudinal deformation is initiated, it will be taken up in the first joint. If the first joint does not have the extension capacity, it is passed along to the second joint, and so on. Strain build-up is minimised which then controls longitudinal loads across pipe joints.'[22]

3.2.1.1. Restrainers

Restrainers are available for PVC and ductile iron RRJ and consist of an extra element bridging the joint. The restrainers consist of a split flange ring with a serrated inside bore that grips the pipe on assembly.

A special flange is fitted to the socket ended pipe and is connected by restraining bolts to the split flange ring on the spigot pipe.

3.2.2. Mechanical couplings

Mechanical couplings can be used on metallic pipes, some plastic pipes and AC pipes. Generally, mechanical couplings as standard are unrestrained. Mechanical couplings can be used on new pipeline construction to provide flexibility with a reliable joint. Couplings will generally provide more flexibility than an RRJ connection. They are also used to facilitate removal of pipes and valves, e.g. at pump stations. Buried pipework allows no axial movement of pipes, even when flanged joints are unbolted. The mechanical coupling or flange adaptor allows pipes and valves to be removed. Another use for mechanical couplings is that of allowing movement (angular deflection) at valve chambers where the pipe is built into a large heavy structure. Mechanical couplings are also used on pipes in tunnels, mainly for construction reasons. Here, pipes are installed from one end, one at a time.

Repairs on water pipelines are often carried out by mechanical couplings. Experienced field staff are vital when installing couplings to guarantee a leak free repair, as backfilling of the excavation may have to be done before recharging while the excavator is present at the repair site.

Mechanical couplings are available in stepped form which allows pipes of different fixed diameters to be joined together. The two diameters are specific for each coupling. These are useful for coupling standard pipe to non-standard pipe or pipes of differing materials, e.g. steel to AC. In the larger diameters, a calculation should be made to determine the axial resultant force which may be considerable if the pressure is high. It may be necessary to install a harness across the joint or anchor blocks to provide restraint.

Up to 600 mm diameter, there is a steel sleeve coupling available which has two tapered sealing rings. This type can accommodate a range of diameters within 20 mm or so of the nominal diameter. These are more adaptable than the step couplings.

3.2.2.1. Unrestrained couplings
3.2.2.1.1. Gibault joint

The Gibault joint (GJ) consists of two cast iron follower rings with a middle sleeve of cast iron or cast steel, two natural rubber sealing rings and pull up bolts (Fig. 12). GJs are also now being fabricated from steel and these are a more reliable coupling than the cast iron

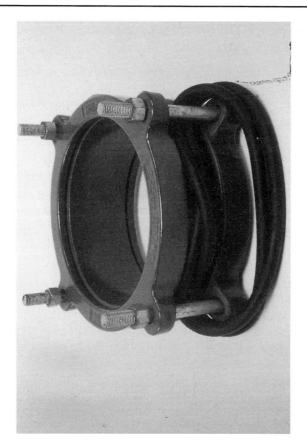

Fig. 12. Gibault joint (courtesy Humes Castings, New Zealand)

versions. When the assembly is pulled up tight on two pipes, the rubber rings are constrained to form watertight seals between the follower rings and the middle sleeve.

GJs are supplied with standard or elongated middle sleeves. The elongated sleeve allows for greater expansion and contraction. GJs are commonly used on steel, cast iron, ductile iron, uPVC, and AC pipes. They are available in various types.

(*a*) *Stepped.* These are for pipes of slightly different diameters.

(*b*) *Tapped branch.* These have a female thread for connecting a small diameter pipe and would more likely be seen on water reticulation than bulk water duty.

(*c*) *Dead end.* These are useful for terminating pipelines and for pressure testing. The dead end must be 'tommed' to resist the axial load.

(*d*) *Flange adaptor.* This fitting allows demounting of a flange pair, e.g. valve removal (refer also to 3.7).

The main advantage of the GJ is cheapness and ease of installation (few bolts). It is possible to crack the cast iron flange rings by incorrect bolt tightening. The recommended angular deflection is low, typically ±1–2 degrees in the larger sizes (750 mm to 1800 mm), but higher in the smaller diameters. Failure of GJs seems to occur because soil settlement has stressed the joint and cracked the follower rings. The rubber seal then relaxes and weeps a fine mist which eventually cuts the rubber ring in half. Undetected, the leak can cut the cast iron follower ring and even the parent pipe where it engages the coupling.

For high pressure applications, steel couplings provide more reliability. GJs are more reliable on low pressure lines, e.g. water reticulation pipes, wellfields or scour valve tail pipes.

3.2.2.1.2. Sleeve couplings

Sleeve couplings are a development of the GJ except that they are made of fabricated steel and the elastomeric sealing ring section is vee shaped, as opposed to a circle (Fig. 13). They also employ more

Fig. 13. Sleeve coupling/flange adaptor (courtesy SPF, UK)

bolts. The pressure rating of sleeve couplings is high, typically 16 bar in the larger diameters, and the standard offering will cope with most situations. Typical angular deflection is ±6 degrees up to 450 mm diameter and ±3 degrees 750 mm to 1200 mm. Sleeve couplings provide some 10 mm movement for thermal expansion, without leaking. This is achieved by deflection of the elastomeric sealing rings, which do not move against the pipe. They are more economical in the larger sizes and offer a reliable jointing system. For example, a 40 year old, 750 mm diameter, 12 km long pipeline was managed by the author. Each pipe length was connected with these couplings and there were no recorded instances of leakage.

Sleeve couplings can be adapted to take up some out-of-roundness and even flats on the pipe, in this case through the use of thicker, i.e. larger than standard cross-section, elastomeric sealing rings. The tightening sequence of the bolts is vital to the proper sealing of the coupling, especially when the fit of the joint is poor.

Sleeve couplings are available in stepped versions, e.g. 972/914 mm and variable fit versions, e.g. 332–349 mm either end. Variable fit versions are now available up to 600 mm diameter. There are two advantages over the standard version: they slide over the pipe in one piece and do not therefore require dismantling; they accommodate more out-of-roundness variation on the pipes to be joined.

The standard AWWA C219-91 governs the manufacture of sleeve couplings. Follower rings are rolled steel angle, flash butt welded and expanded to the final size (diameter). Similarly, the sleeves are flash butt welded rolled steel, expanded to the final size. Elastomeric sealing rings are made of EPDM or natural rubber. There are sleeve couplings on the market with the follower ring angle section fabricated from two plates, and hand fillet welded. Similarly, the middle sleeves are hand welded. The standard sealing rings are natural rubber, not EPDM.

Sleeve couplings are available with welded pillars on the middle sleeve and follower rings to allow cathodic protection bonds or bond cables to be connected. This integrates the coupling into the rest of the pipeline for the cathodic protection system.

Insulated couplings are available which work by providing a skirt (land) on the elastomeric sealing ring, thus preventing the inside of the follower ring from contacting the pipe. A rubber insulator is also provided to keep the two ends of the pipe from making contact (refer chapter 6). One manufacturer offers a non-conducting neoprene gasket which can also be retrofitted to existing couplings.

Fig. 14. Lip seal coupling (courtesy Straub, Switzerland)

Finally, sleeve couplings can be installed with a harness in areas subject to seismic activity. The large tie-bolts (e.g. half the cross-sectional area of the pipe) installed across the joint are designed to provide a controlled yield before the joint pulls apart. In this way, damage to the pipeline should be minimised in a seismic event.

3.2.2.1.3. Lip seal couplings

The lip seal coupling consists of a casing made of stainless steel, housing a circular rubber lip seal and captive bolts to tighten up the coupling (Fig. 14). The assembly is slid over the pipes to be joined, and a small gap is left between pipes before the bolts are tightened to a pre-set torque. The internal pressure of the water forces the lip seal against the pipe and the casing and provides a sealing force. The lower pressure rating couplings have stainless steel 304 grade casings. The higher pressure couplings, typically 10 bar upwards, use hot dip galvanised steel for the casing. Advantages of this type of coupling are fewer bolts, hence simpler and quicker installation. Also, they are a better profile to apply a protective coating to (refer 3.2.2.6).

Finally, out-of-round pipes within reason can be joined with these couplings where no other sealing solution is possible.

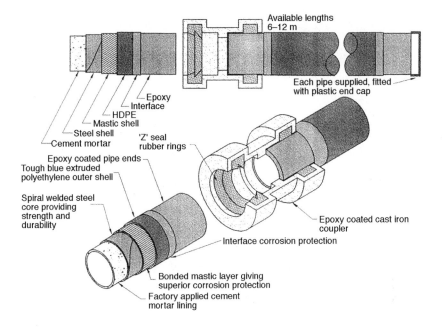

Available lengths
6–12 m

Each pipe supplied, fitted
with plastic end cap

Epoxy
Interface
HDPE
Mastic shell
Steel shell
Cement mortar

'Z' seal
rubber rings

Epoxy coated pipe ends
Tough blue extruded
polyethylene outer shell

Spiral welded steel
core providing
strength and
durability

Epoxy coated cast iron
coupler

Interface corrosion protection

Bonded mastic layer giving
superior corrosion protection
Factory applied cement
mortar lining

Fig. 15. Epoxy coated cast iron coupler (courtesy Humes Steelpipe Ltd, New Zealand)

3.2.2.1.4. Socket couplings

Socket couplings are used to couple smaller diameter pipelines having plain ends (refer Fig. 15). They consist of a cast iron socket with a central locating stop and with two elastomeric sealing rings located in grooves (Fig. 15).

3.2.2.2. Restrained couplings

The use of restrained couplings avoids the use of anchor blocks at bends, tees and dead ends. Small axial movements due to temperature fluctuations can be accommodated to a certain extent.

3.2.2.2.1. Lip seal couplings

The restrained version of the lip seal coupling is virtually identical to the standard unrestrained version, but has in addition two serrated metal rings built into the casing ends, which indent into the pipe to provide restraint. These couplings are easy and quick to install, and it is easier to apply protective wrapping to their profile than to a sleeve coupling.

The coupling is also highly resistant to shock loadings and to high levels of angular deflection without leaking.

3.2.2.2.2. Sleeve couplings

This sleeve coupling is similar in design and construction to the sleeve coupling described in 3.2.2.1.2. The sealing rings, however, have load bearing metal inserts which grip the wall of the pipes. Internal pressure causes the assembly to lock further, thus providing full axial restraint. These couplings are available to connect ductile iron pipe to BS 4772 up to 250 mm nominal diameter and to connect steel pipes up to 200 mm nominal diameter. Working pressures are 16 bar up to and including 200 mm nominal diameter and 10 bar in 250 mm nominal diameter.

3.2.2.2.3. Two-piece type couplings

Two-piece type couplings are designed for the American National Standards Institute (ANSI) diameter pipes and therefore compatibility with other pipelines should be checked. The standard material is ductile iron. They are also available in stainless steel and aluminium alloy.

3.2.2.2.3.1. For plain ended pipe.

These couplings are available up to 400 mm diameter at 10 bar pressure rating. The C-shaped gasket seals across the joint, being held in position by a two-piece coupling bolted together. Tightening of the bolts causes hardened steel jaws built into the coupling to indent into the pipe wall to provide restraint.

3.2.2.2.3.2. For grooved pipe

Couplings are available up to 1050 mm diameter at 10 bar pressure. These fit grooved end pipe. It is possible to produce these ends on concrete-lined steel pipe in the workshop and the field. Grooves can be made in pipe ends by rolling or cutting if the pipe has sufficient wall thickness. The coupling 'feet' sit in the grooves on each pipe end. This mechanism provides axial restraint. Couplings can be specified to allow flexibility, i.e. expansion/contraction and angular deflection or rigidity.

3.2.2.3. Repair clamps

Some types of repair clamp, e.g. repair mufflers, can be fitted over a leaking joint or pipe under pressure, thus avoiding a (costly) shutdown. The leaking joint is then totally encapsulated within the repair clamp. Repair clamps are well suited for repairing leaks on pipes in tunnels with limited room for working. Repair clamps have the advantage of fitting over the pipe in situ. This minimises any possibility of external

Fig. 16. Repair clamp for cast iron joints (courtesy SPF, UK)

contamination.[23] The alternative repair method of using two mechanical joints and a section of pipe requires a larger excavation, and cutting the pipe in two places creates the real possibility of external contamination (refer 12.2.2).

3.2.2.3.1. For spigot and socket joints
Split design repair clamps are available to connect across the bell of cast iron pipes with leaking lead joints (Fig. 16). They are expensive so it is usually cheaper in materials to cut out the offending section and splice in a new pipe with two couplings. However, the cost and ramifications of a shutdown for repair can justify the use of these couplings.

3.2.2.3.2. Split clamps
Split stainless steel clamps in 304 and 316 grade are available. They consist of a casing with a honeycomb elastomeric seal bonded to the

inside of the casing and a number of bolts to pull the casing together around the leaking pipe, thus sealing the leak. Split clamps are also available in ductile iron with a 'waffle' seal. Clamps can be used to join pipes with dissimilar outside diameters and they can also accommodate some out-of-roundness.

Clamps are designed to get you out of a fix. Conceptually, a repair clamp is only for repairs. It cannot be considered a permanent solution unless the engineering and construction of the clamp reflects this. The ability of the clamp to deflect axially and to retain sealing under full working pressure is the critical parameter.

One manufacturer states that their stainless steel clamp can be used for normal pipe jointing. This scenario is unlikely as the cost of the clamp is considerably more than a standard sleeve coupling.

Clamps with tee branches built in can be used on cast iron or AC mains as an alternative to splicing in a section of pipe with a fabricated branch. Their use on high pressure, large diameter pipelines is debatable.

3.2.2.3.3. Lip seal repair clamps

Lip seal repair clamps are similar to the standard coupling but have radial slots in the stainless steel casing to allow fitment around the pipe (Fig. 17). The higher pressure galvanised steel casings have a

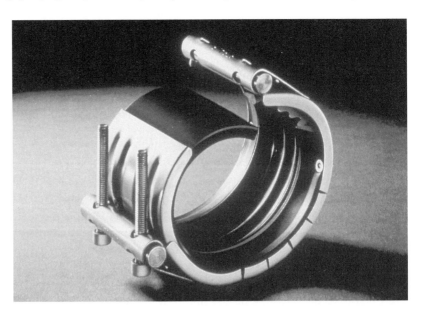

Fig. 17. Lip seal repair clamp (courtesy Straub, Switzerland)

hinge to allow fitment around the pipe. Axial movement in the pipeline is not needed to fit these clamps as they wrap around the pipe leak. They are very useful when broken Gibault joints are being replaced, as no cutting of the pipe is required. This makes for a quick repair. This type of repair clamp provides the same axial deflection as the normal coupling.

The author has used lip seal repair clamps to replace corroded sleeve couplings on a pipe cantilevered from the side of a road bridge spanning an estuary. From a floating platform, the existing couplings were removed with a grinder and lip seal repair clamps were installed in situ, 300 mm diameter and 20 bar (200 m) pressure rating.

3.2.2.4. Asbestos cement socket couplings
Asbestos cement socket couplings consist of a loose sleeve of asbestos cement with two elastomeric sealing rings. The Widnes joint fits plain ended pipe, while the Comete joint requires pre-machined pipe ends. The two types are not compatible.[24]

3.2.2.5. Glassfibre reinforced polyester socket couplings
The glassfibre reinforced polyester (GRP) socket couplings are a collar with a full width EPDM lip seal. A special variety provides full axial restraint.[25]

3.2.2.6. Corrosion protection of metallic couplings
All metallic couplings should be specified with a shop-applied protective coating. In addition to this, a field-applied coating system should be used, e.g. 'densotape'. The coupling bolts must also be protected, as these can waste in corrosive soils and lead eventually to joint failure.

3.3. Flanged joints
A flanged joint is a combination of a mechanical joint and a welded joint (Fig. 18). The flanged joint provides full axial restraint so that it can be used on dead ends, often for pressure testing a new pipeline. Another application for flanges is on smaller diameter pipelines having a steep grade, e.g. rising branchlines. Headwalls to anchor the pipe are not needed if flanges are used.

Blank (blind) flanges in the larger diameters are often strengthened by flat bars (strongbacks) welded on to the back of the plate. This allows the use of plate that is thinner than unstrengthened blank flanges which are typically up to 400 mm diameter and found on pipework in

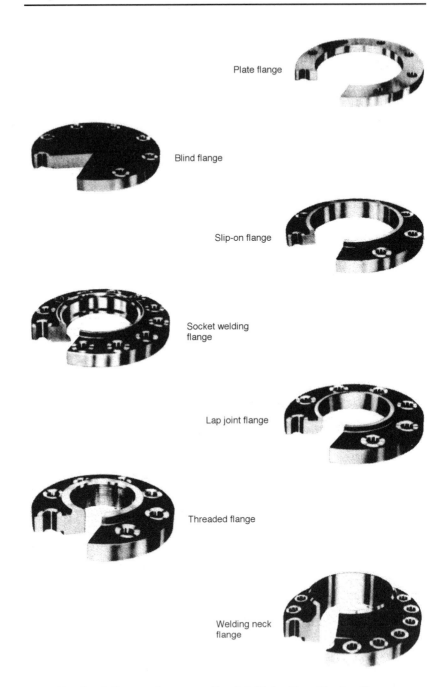

Plate flange

Blind flange

Slip-on flange

Socket welding flange

Lap joint flange

Threaded flange

Welding neck flange

Fig. 18. Steel flanges (courtesy Steel & Tube New Zealand)

the process industries. Blank flanges are fitted with 'blank' gaskets to prevent the potable water from contacting the underside of the steel blank flange.

Flanges can be used on bends to avoid the use of anchor blocks. This application avoids the use of gravity anchor blocks — blocks on lines with a bend causing a resultant upward thrust. Flange pairs also provide a rigid joint for pipes in a self-supporting situation, e.g. stream crossings. Flanges may also be used on under stream crossings where the pipe is later buried in concrete. The flanges facilitate easy rigging and installation of pipe in the smaller diameters.

Flanged joints can leak and the rubber gasket blow out. This can occur even if the pipe is buried in concrete. Correct tightening of bolts when the joint is being made is important. Bolts should be snug tightened, then fully tightened using a pattern of opposite bolts.

The use of flanges on pipelines subject to dynamic loadings or bending loads caused by soil settlement is debatable.

3.3.1. Pressure rating

According to the Standards, the pressure rating of flanges is usually described as the non-shock maximum working pressure. The common flanges are shown in Table 2.

3.3.2. Slip-on flanges

The slip-on flange slips over the steel pipe because it has an inside diameter 3 mm or so greater than the outside diameter of the pipe. When flanges are being ordered, it is imperative to specify the inside bore to suit the pipe diameter being used, e.g. NZS 4442, ANSI, etc. The flange is installed about 6 mm proud of the end of the pipe to allow a seal weld (typically 3 mm fillet) between the pipe wall and the flange bore. A larger fillet weld, sometimes referred to as a strength weld, is made to connect the back of the flange to the pipe wall. The fillet leg length should be a minimum of the pipe wall thickness. Slip-on flanges can be made from structural grade steel plate or supplied as a forging. Forgings are intrinsically strong although they are more expensive. Carbon steel flanges weld easily to steel water pipes whether they are plate or forged type.

3.3.3. Blank (blind) flanges

The blank flange can be supplied in plate or forged steel. They are used on spare branches so that, in future, a flanged valve can be fitted easily and the branch pipework installed away from the valve. If a valve is

Table 2. Pressure rating of British and American flanges

BS 10	Pressure, m	Pressure, psi
Table D	70	100
Table E	140	200
Table F	210	300
BS 4504	**Pressure, m**	**Pressure, psi**
PN6	60	88
PN10	100	147
PN16	160	235
PN25	250	368
AWWA/ANSI	**Pressure, m**	**Pressure, psi**
Class D/125 lb up to 300 mm	122	175
Class D/125 lb 350 mm above	105	150
Class E/250 lb	192	275
Class E/150 lb	192	275
Class E/300 lb	504	720

Note 1: ANSI 125 lb and 150 lb have identical drillings.
Note 2: ANSI 250 lb and 300 lb have identical drillings.

fitted on a spare flanged branch, a blank flange should be fitted to the valve to prevent leakage if the valve is inadvertently opened. As noted in the introductory section, large blanks can be ribbed at the back for economy.

3.3.4. Weld neck flanges

The weld neck flange is supplied as a forging which butt welds to the pipe. It is used on high pressure duties, supplied with a raised face, and for very high pressures supplied with a ring joint connection.

3.3.5. Screwed flanges

3.3.5.1. Small diameter flanges

Small diameter flanges are threaded on the inside bore to screw on to threaded pipe (galvanised or stainless steel). They allow flanged valves to be fitted in a screwed system. Screwed flanges are typically small diameter up to 100 mm.

3.3.5.2. Ductile iron flanges

Ductile iron flanges are screwed on to the pipe wall to provide a flanged connection for flanges and valves.

3.3.6. Flanges for plastic pipes

The flange used for connecting uPVC plastic pipes corresponds to a socket weld flange used on metallic pipework. The pipe fits into the socket of the flange and the 'weld' is made with solvent cement. A similar flange is used that consists of a solvent weld 'stub end' and a galvanised backing flange. This is because the plastic alone cannot take the bending stress in the larger sizes and pressure ratings. PE flanges can be fusion welded on to PE pipe.

Blank flanges for plastic pipes can be galvanised steel blank flanges fitted with blank gaskets so that the water does not contact the steel.

3.4. Flange faces and gaskets

Flange faces on cast iron valves and fittings should be flat faced. This is to prevent undue bending stress on the cast iron flanges. Steel flanges can be supplied with flat or raised faces. The common gasket material is natural rubber with reinforcing strands in the body of the rubber. Rubber gaskets (3 mm thick) are normally full face for flat faced flanges. They are easily made in the field by tapping a hammer on a sheet of rubber laid over the flange, around the flange profile and holes.

Raised face flanges can be used on water lines. The non-asbestos gasket sits inside the bolt circle and looks like a ring, although full face rubber gaskets can also be used on raised face valves and flanges.

3.5. Bolts

Hot dip galvanised ordinary grade bolts (250 MPA yield) are used on flanges at normal pressures. Black bolts corrode far more quickly and, consequently, should not be used. Imperial bolts are not always available in a hot dip galvanised finish and metric bolts can fit the imperial flange standard of BS 10 in most applications.

Isolating flanges on cathodic protection (CP) systems sometimes require the use of black imperial high strength (640 MPA yield) bolts. This is because the bolt has to fit inside a plastic (non conducting) sleeve, fitted inside the flange bolt holes. The bolt sometimes has to be two diameters below standard in order to fit. The loss in bolt area would also mean a loss in bolt tension and a likely gasket leak if high strength bolts were not fitted. Metric high strength hot dip galvanised bolts (for structural steel) will sometimes fit the CP flange. When a metric bolt will not fit, a high strength black bolt has to be used. This last point leads naturally to protection of flanged joints and bolts.

3.6. Protection of metallic flanged joints

Flanged joints can be protected if an epoxy paint is applied to the clean dry surfaces. Another method for buried flanges is the application of a petrolatum tape system. This protects the whole of the joint and also the bolts, and allows easy dismantling of the joint in the future, if required.

3.7. Flange adaptors

Flange adaptors are a combination of a flange and a half sleeve coupling. They are used in tight situations providing axial movement to facilitate the removal of valves and pipework. They are economic as they are equivalent to a flange and a coupling. Without such an arrangement it is impossible to replace a leaking flange gasket. Without an adaptor, the pipe would have to be cut to allow movement and then rewelded afterwards.

All significant valves should be fitted with flange adaptors or their equivalent. Flange adaptors are provided with factory protective coatings. Further protection, such as petrolatum tape, is desirable.

3.8. Lead packed joints

Cast iron pipes of the spigot and socket variety were often joined together by means of pouring molten lead into the joint and caulking it afterwards. These joints can still be made with lead in the traditional way although modern safety considerations may rule this out. Failure of lead joints is the result of lead being forced out of the socket. Lead joints can be mechanically re-caulked and a joint harness installed hard up against the socket in order to restrain the lead. Special repair couplings are available (refer 3.2.2.3.1). An alternative solution is to install lip seals inside the pipe, if there is sufficient access.[26] The pipe interior across each joint must be cleaned before the lip seal is installed, spread with an hydraulic spreading tool and, finally, pressure tested.

4. Pipe fittings

If a pipeline system is conceived as a giant construction set, then pipe fittings are the standard pieces along with the pipe sections themselves. One great benefit of pipe fittings is standardisation. Pipe fittings are generally available in metallic and non-metallic materials. I have not covered every pipe fitting because some will rarely be used and are more likely to be seen on water reticulation, rather than bulk water duties.

The common approach in steel water pipelines (lined) is to fabricate the 'fittings', either in the workshop or on site. Such flexibility is not available with many other materials.

4.1. Pipe ends
4.1.1. Plain end
The plain end pipe fitting is designed to fit into a joint, either RRJ or mechanical coupling.

4.1.2. Bevel end
The bevel end is designed to butt weld to another element. Unlined steel pipe fittings with bevel ends are available in ANSI B16.9 (BS 1640) and BS 1965, for example. These fittings if hand-lined, or even used unlined, can be used as alternative compatible materials in an emergency situation. Stainless steel fittings to ANSI B16.9 can be used on small diameter lines (refer 1.5.2) or larger diameter lines for stream crossings (refer 1.5).

4.1.3. Socket end
Socket ends allow pipe elements to be connected inside the socket of the fitting, e.g. RRJ connection. Similarly, some plastic systems use the socket end for solvent welding.

4.1.4. Threaded end

Threaded ends can be furnished male or female. Typical fittings are small diameter galvanised malleable iron to BS 143, stainless steel 150 pound fittings fashioned on BS 143, BS 1740 steel and ANSI B16.11 (BS 3799) in steel or stainless steel.

4.1.5. Flanged end

Flanged end pipe fittings have the advantage of providing a flanged joint able to resist end loads (refer 3.3).

4.1.6. Grooved end

Grooved end fittings are available in galvanised steel, stainless steel and aluminium (refer 3.2.2.2.3.2).

4.2. Types of fitting

4.2.1. Bend (elbow)

Two parameters define the elbow, the degrees and the radius. Typical elbows are 45° and 90°. Radii are long or short. Bends can have support feet (duckfoot) incorporated to provide a support. These are used in above ground situations, e.g. pump stations.

4.2.2. Knuckle joint

Knuckle joints, supplied in ductile iron, provide up to 15° deflection either direction. They are used for under river crossings to provide flexibility in the pipeline. Two knuckle joints with an expanding sleeved section of pipe between will provide good articulation in a seismic situation (Fig. 19).

4.2.3. Tee (branch)

Tees are either equal or unequal. Equal tees have three identical ends. Unequal tees have one smaller diameter connection, usually on the branch, but occasionally on the main. Fire hydrant tees are a typical example of an unequal tee.

4.2.4. Taper

Tapers are cone shaped and are used as a transition piece between pipes of different diameters (Fig. 20). The angle of the taper is typically 8° which results in little head loss, used in either direction. Tapers can be supplied concentric or eccentric. Concentric tapers keep the two joining pipes coaxial, while the eccentric taper keeps the top of the pipes flat, or the bottom of the pipes flat. Typical applications for

Fig. 19. Double knuckle joint (courtesy SPF, UK)

tapers are on main valve installations, where a smaller diameter valve is often used to save money and the small head loss of two tapers is insignificant. Another application is on pump suction lines which usually are a size larger than the pump suction flange connection.

Fig. 20. Taper (courtesy Wellington Regional Council, New Zealand)

Here, as in the main valve application, top flat eccentric tapers are used so that no air lock occurs.[27]

4.2.5. Reducer

Reducers, supplied under ANSI B16.9 (BS 1640), can be used in either direction. The cone angle is much greater than a taper, with corresponding higher head loss. They can be supplied concentric or eccentric. Reducers are economical fittings to use on scour lines where the scour pipe size is larger than the scour valve, to reduce the scour velocity.

4.2.6. Reducing flange

Reducing flanges are supplied in ductile iron and adapt flanged pipes of different diameters. This is achieved by machine screws 'bolting' the small diameter pipe flange to the reducing flange, which has tapped holes. The large diameter pipe flange bolts to the reducing flange using conventional bolts.

4.2.7. Socket (straight coupling)

A socket is a hollow bar with female threaded ends or socket weld ends to connect two pipes together. These fittings are small diameter (refer 4.1.3 and 4.1.4). Larger diameter sockets are available specifically for water pipe and weld on the outside only, like a slip joint (refer 3.1.3). Epoxy mortar has to be put on the inside of the joint before assembly and welding take place on these pipes, which are too small to access for lining repairs.

4.2.8. Union

A union, like a socket, can be female threaded or socket ended. They are used in small diameter pipe systems to allow easy dismantling of pipework (refer 4.1.3 and 4.1.4 for Standards). The union has three parts, namely two tails with an outside thread and a nut which pulls the machined faces of the two tails together to form a seal. Some unions have a gasket between the machined faces.

4.2.9. Bush

The main application for bushes is that of adapting small diameter pipes to pressure gauges and other small equipment. Bushes are screwed transition couplings. The large end is male and the small end is female.

Similar couplings exist in plastic having socket ends for solvent welding.

4.2.10. Cap

Caps are a means of blanking off the end of a pipe. They can be welded (socket or butt) or screwed on.

4.2.11. Plug

Plugs are a means of blanking off a pipe fitting. A typical application is the repair of leaks (refer 1.4).

5. Valves

Valves in a bulk water pipeline system control the flow of water. The smooth operation and tight closing of the valves are the critical aspects. Large valves should be housed in valve chambers providing easy access for maintenance or future replacement. The chamber also acts as a thrust block to take the forces when the valve is shut. Buried valves with leaking glands need to be dug up to gain access to the valve. Chambers should be installed for all valves where practically possible, as the cost of digging up valves later, exceeds the cost of a chamber at the time of construction.

Gearboxes are required on some valves for one of two reasons

(a) to control the rate of opening and closing, e.g. main valves or plug valves on buried duty; this avoids surges and water hammer
(b) to overcome high seating pressures, e.g. scour valves which have high differential pressure across them.

Gearboxes should be fully enclosed and waterproof for buried service. It is worth checking the flanged connections on the valve gearbox drive to ensure there is a sealing gasket (plug and butterfly valves). Older sluice valve designs had no box and a vertical pipe sleeve was installed to keep dirt out of the gears and to provide access. This arrangement was only partially successful, as high groundwater flows mud into the valve operation access shaft.

All valves in the system should preferably be exercised *once a year*[28] to check the smoothness of operation and tight shut off (ascertained by listening on the valve key). The status of the valve, i.e. open or closed, can be checked at the same time, as valves can be inadvertently closed or opened. Manual operation of valves using an extension (key) and bar is standard practice. Hydraulically powered valve actuators are available, with adjustable torque limiters to prevent overtightening of the valve. Failures in valves such as broken spindles seem to stem from lack of operation. Gland leaks, if left, can literally machine the valve spindle so that a replacement is required. Re-packing of sluice valve glands can be carried out under pressure on some valve designs. On others, a shutdown and depressurising of the pipeline is required. Corrosion of bolts on a buried valve can cause a leakage failure. Valves require corrosion protection (refer 3.6). Again, a chamber allows for easy inspection and maintenance. Valves can often be reconditioned for between one-third and one-half of the replacement cost.

The rotation direction of operation of all valves should be the same, to reduce the possibility of valve spindles being broken by operators trying to open further an open valve and to shut further a shut valve.

Marker posts with valve descriptions and distances to the valves are invaluable when particular valves are being sought. These can be made of wood and concreted into their surroundings (in open areas). Where vandalism is a problem, concrete posts can be used. Sealing over of valve blocks in roads is a continuing problem best addressed by asking roads authorities for their resealing programme and by developing a good relationship with the local roads Engineer(s). The yearly inspection of valves will pick up these non-conformances.

Valve materials
The body parts of valves operating up to 25 bar (250 m) are typically cast or ductile iron. Pipeline systems with pressures above 25 bar employ valves of cast steel construction, for strength reasons. These are not discussed further. Typical valve materials are shown in Table 3. For exact details, refer to the specific valve standard.

5.1. Main (line) valves
The function of main valves is to section the pipeline into manageable lengths. This is apparent when working on a section of main, e.g. to inspect it internally or to replace a main valve. This latter activity requires two sections of the line to be shut and drained. The constraints of the customer's reservoir storage time determines the time available to work on the main. The author has managed a shutdown requiring the draining of three sections of 900 mm diameter main within a 20 hour period (2 peak demands), which was within the storage capacity of the supplied reservoirs. Four sections of main drained, worked on

Table 3. Typical valve materials

Valve part	Material
Body, bonnet, stuffing box	Cast iron or ductile iron
Stem seal thrust housing, bridge	Cast iron or ductile iron
Stem cap, handwheel (if fitted)	Cast iron or ductile iron
Spindle or shaft	Stainless steel, copper alloy
Primary moving element, e.g. gate, plug or disc	Cast iron or stainless steel
Gate or plug lining	Nitrile rubber
Valve body lining	Epoxy, fusion bonded thermoplastic
Valve body resilient lining	Nitrile rubber or EPDM
Valve seat	Nickel, copper alloy

and recharged would probably be too much. Main valves approximately 5 km apart allow for draining, internal inspection (if large enough) and recharging within the parameter of a day's storage.

Other pipeline systems may have different parameters. The water system should allow for replacement of *any* main valve without compromising supply. There are methods of replacing main valves *without draining*, such as creating an ice block either side of the valve using liquid nitrogen.[29] Other methods allow a live tap to be carried out on large diameter pipes. This is expensive, but has no peer in critical situations where large mains *cannot* be shut down.

Older main valves are probably sluice valves. Sluice valves (Fig. 21), if operated regularly, are generally reliable, as they are of rugged construction. Overtightening of these valves can break or bend the spindle. This is best addressed through training of the operator. Valves should be 'backed off' half a turn or so to indicate the valves' position for the next operator. Then, if the operator turns the key the wrong way there is some easy turning (half a turn) before the end position is reached. Sluice valves are tall. If the pipeline has little cover the valve may stick out above grade. Replacement with a butterfly valve can solve this 'problem' as the valve is more compact. Tall sluice valves may have been installed horizontally with bevel gearing in order to avoid this height problem.

The modern choice for main valves 400 mm or so up is the butterfly valve or plug valve, because it is cheaper, compact and light (Fig. 22). Butterfly valves are supplied in various end configurations. The double flanged end is preferred as it allows the removal of pipework on the depressurised side, which is essential when large diameter pipes are being inspected. Butterfly valves need to be specified to take the full

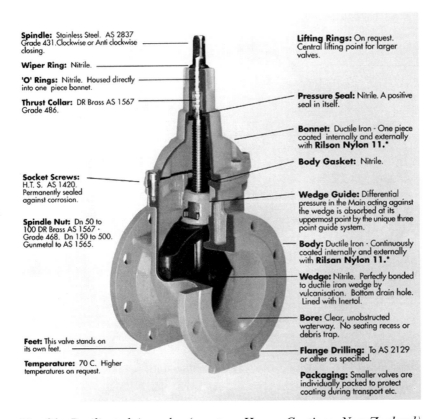

Spindle: Stainless Steel. AS 2837 Grade 431. Clockwise or Anti clockwise closing.

Wiper Ring: Nitrile.

'O' Rings: Nitrile. Housed directly into one piece bonnet.

Thrust Collar: DR Brass AS 1567 Grade 486.

Socket Screws: H.T.S. AS 1420. Permanently sealed against corrosion.

Spindle Nut: Dn 50 to 100 DR Brass AS 1567 - Grade 468. Dn 150 to 500. Gunmetal to AS 1565.

Feet: This valve stands on its own feet.

Temperature: 70 C. Higher temperatures on request.

Lifting Rings: On request. Central lifting point for larger valves.

Pressure Seal: Nitrile. A positive seal in itself.

Bonnet: Ductile Iron - One piece coated internally and externally with **Rilson Nylon 11.**

Body Gasket: Nitrile.

Wedge Guide: Differential pressure in the Main acting against the wedge is absorbed at its uppermost point by the unique three point guide system.

Body: Ductile Iron - Continuously coated internally and externally with **Rilsan Nylon 11.**

Wedge: Nitrile. Perfectly bonded to ductile iron wedge by vulcanisation. Bottom drain hole. Lined with Inertol.

Bore: Clear, unobstructed waterway. No seating recess or debris trap.

Flange Drilling: To AS 2129 or other as specified.

Packaging: Smaller valves are individually packed to protect coating during transport etc.

Fig. 21. Resilient sluice valve (courtesy Humes Castings, New Zealand)

differential pressure across the valve. (Assume one side of the valve is drained for inspection.)

'If a large number of butterfly valves are used, pressure loss in a distribution system can be significant. Another disadvantage is that butterfly valves prevent the use of pigging which involves using line pressure to force a bullet-shaped plug through a water line to locate and clean flow restrictions.'[30] Some types of butterfly valve need to be open when installed as the bolting up of the valves is an integral part of the sealing of the disc to the body. If installed in the closed position, the valve if opened and subsequently closed will not seal properly.

If a butterfly valve is overtightened, this will cause the shear pin to fail so that the spindle will rotate freely without actuating the valve. This mechanism protects the valve from damage.

Plug valves are another choice but they have pressure limitations, especially in the larger sizes (Fig. 23). Spectacle blinds may be required

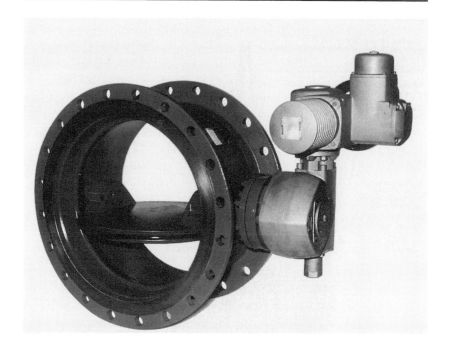

Fig. 22. Butterfly valve (courtesy Jansen Armaturen, Germany)

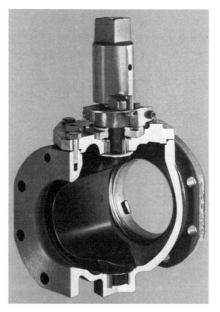

Fig. 23. Cam-centric valve (courtesy Valmatic, USA)

in addition to main valves in order to provide more safety during internal inspections. A simpler method of providing isolation *and* access for internal inspection at main valves is to have two removable sections of pipe, effectively constituting dismantling joints, either side of the valve. This arrangement allows physical blanking of the shut valve and provides easy access to the pipe either side of the valve (refer Fig. 24). Isolation of the bypass arrangement is achieved by the removal of a flanged bend and the bolting of a blank flange to the shut bypass valve.

5.1.1. Electrical operation of main valves

Main valves can be electrically actuated for local or remote operation. It should be noted that manual operation of main valves is more

Fig. 24. Dismantling joints on main valve (courtesy Wellington Regional Council, New Zealand)

foolproof and, anyway, the shutting of a section of main for maintenance purposes requires the services of several people.

The speed of the actuation is the critical parameter. On a large valve, the actuation speed must be slow to avoid surges. The instantaneous closing time has been defined as $2l/c$ or less, where l is the length of the line between the valve and the exit point, or the inlet point and the valve, and c is the speed of sound in water. The end of the 'rapid envelope' is $20l/c$.[31] On one problem valve the author encountered, valve closing time was 125 seconds with an instantaneous closing time calculated at 50 seconds. The valve was thus operating in the rapid closing envelope. Large surges in the pipeline shattered an old cast iron main that normally operated comfortably around 30 m head.

The actuator was controlled from a control panel in the treatment plant. The cause of the problem was a temporary signal dropout between the control panel and the actuator which caused a 6 turn close on the valve (50 turn total). This was occurring with the valve half open. Surge problems can therefore occur not just in the last 15% of valve closing/opening. The fault was rectified by interposing a 6:1 gearbox between the actuator and the valve gearbox. This gave mechanical security regardless of the control panel signal. The closing time was then 720 seconds which was outside the 500 seconds rapid closing envelope. The valve can now be shut at full actuation speed without surge problems.

One actuator manual suggests that large valves subject to the possibility of surges can be actuated in short bursts of the actuator. This is not intrinsically foolproof as the rate of close/open of the standard valve may be too high. Any electrical fault that causes the valve to open/close several turns at a time may well cause surge problems and damage. These valves can be specified with a gearbox which takes the closing time outside the calculated rapid closing time. The actuator will also run better and require less maintenance if it does several turns at a time, rather than fractions of a turn.

Another solution is to time the valve to allow a multiple phase closing sequence. 'At three of the four control valve sites, the Figure 1000 valves close from 100% open to 15% open in 30 seconds, and then take 80 seconds to reach the closed position...'.[32]

5.1.2. Valve chambers (vaults)

Large valves, e.g. 600 mm diameter and larger, are usually installed in valve chambers to facilitate maintenance and operation (Fig. 25). Valve chambers also act as an anchor block to resist the thrust

Fig. 25. Main valve chamber (courtesy Wellington Regional Council, New Zealand)

(often a large force) when the valve is closed. If large valves are buried, this does not allow access to the glands (sluice valves), although for butterfly valves there are few negative consequences. Buried valves installed with corrosion protection, e.g. petrolatum tape system, should not corrode significantly. The external bolts also need protection to prevent wastage.

Valve chambers may also house branch valves and scour valves either side of the main valve. There may also be control valves, e.g. pressure reducing valves, supplying a zone for emergency duty, acting as a large auxiliary supply.

Valve chambers can be quite large structures with heavy concrete roof slabs supported by structural steelwork. Groundwater is the enemy of valve chambers and seepage through the joints in the structure is usually removed by a sump pump installed in one (low) corner. This is because most large valve chambers are too deep for a gravity drain to work. The dampness in valve chambers can cause fast corrosion of the structural steel. Stitch welding of structural steelwork should be avoided because it encourages crevice corrosion. Hot dip galvanised steelwork provides good corrosion resistance. Badly corroded steel can be sandblasted and repainted, provided that the atmosphere is dried. This may be achievable only if the roof slabs are lifted off. Then, the top bearing surface of the support beams previously covered by the roof slabs, can be maintained.

Access to a valve chamber requires the air to be tested for safety, if the chamber is deep. The chamber probably qualifies as a confined space (refer chapter 13).

Valve chambers may have electrical panels for the lights, panel heater, electric valve actuation, sump pump, electrical sockets to allow maintenance, etc. The electrical panel may also have an alarm and sensor to monitor for a flood condition. Maintenance of the electrical equipment must be carried out by a qualified electrician.

5.1.2.1. Small valve chambers

Small valves up to, say, 300 mm diameter can be installed in simple chambers made from precast concrete risers. A concrete lid is placed on top of the top riser and is located by the rebate on the riser. Typically, the concrete lid has a 600 mm cast iron lid and frame fitted to provide access to the chamber. Stainless steel or plastic protected steel ladder rungs provide a safe descent to the bottom of the chamber (Fig. 26).

Fig. 26. Scour valve chamber (courtesy Wellington Regional Council, New Zealand)

It is difficult to seal the pipe penetration through the chamber's walls and the chamber may have perpetual groundwater covering or part covering the valve. There are special concrete sealing products which can be used to waterproof the seal. Dewatering such chambers is difficult because pumping out the water causes more groundwater to ingress. A solution is to create a dewatering point close by the chamber, consisting of a bed of drainage metal with a pump out riser pipe. This allows remote dewatering of the chamber during initial construction. The dewatering point can also be used at a later date should it be necessary.

5.1.2.2. Access steelwork inside valve chambers

Access steelwork inside valve chambers has to be designed to allow the removal of certain pieces of equipment that may require future maintenance or be designed for easy, quick removal. The width of walkways should be 750 mm minimum. Vertical ladders are typically 500 mm wide with 20 mm rungs at 300 mm centres, plug welded to two stringers made of angle or flat bar. All the steelwork should be hot dip galvanised for corrosion resistance. Walkway floors are typically made of galvanised grating, although aluminium floor grates can be used. Walkways require handrails with a top rail, knee rail and kickflat. The design may have to cater for a valve point load, but more likely not because the valve would be removed and lifted straight out of the valve chamber through a removed roof panel.

5.2. Bypass valves

Bypass valves are fitted on main valves and some branch valves (200 mm up). Their function is to provide a slow charging rate across the main *shut* valve. This avoids surging and allows the air in the section of pipeline being charged to escape by way of the air valves. They are typically 100–150 mm diameter.

5.3. Branch valves

The function of the branch valve is to isolate the branchline from the main pipeline (Fig. 27). The branch valve should be fitted as close as possible to the main pipeline. Usually the branch valve will be flanged. A flanged inlet spool, incorporating the bypass connection, may be fitted between the branch valve and the flanged branch fitted to the side of the main. This spool may be cast iron on older pipelines. If the bypass connection was fitted directly to the branch on the main,

Fig. 27. Branch valve replacement (courtesy Wellington Regional Council, New Zealand)

it would save this spool. Branch valves should be housed in chambers for easy maintenance and repair.

5.4. Air valves

The function of air valves[33–35] is threefold: first, to allow air to escape when the pipeline is being charged; second, to allow air to enter the pipeline when water is being discharged from the pipeline; third, to allow dissolved air to escape at minor high points. The double air valve (Fig. 28) achieves all these functions with two valves of differing orifice areas, connected to one common inlet. The large orifice valve achieves the first two functions and the small orifice valve achieves the third function.

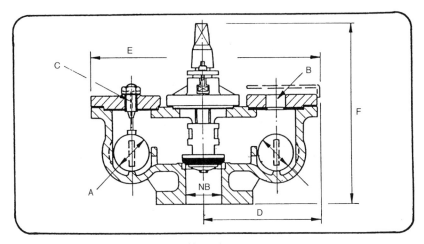

80 mm size

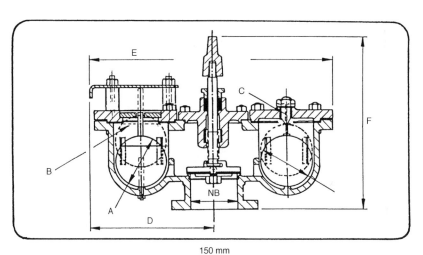

150 mm

Dimensions: mm						
NB	A Ball dia.	B Large outlet dia.	C Small outlet dia.	D	E	F
80	76	42	1·6	255	490	330
150	152	98	1·6	400	790	530

Maximum working pressure: 16 bar

Body test pressure: 24 bar

Seat test pressure: 16 bar

Maximum service temperature: 20°C

Valves are supplied flanged and faced to BS 4504:1989, PN16 Type B - raised face.
Drillings to be specified with order.

Fig. 28. Double air valve (courtesy Gillies Foundry & Eng. Co. Ltd, New Zealand)

Vent-O-Mat. Series RBX Operation

Pre-notes:

1. Venting of a filling pipeline

The operation of a kinetic air release valve is such that fast approaching water is almost instantaneously halted by the valve's closure without the shock cushioning benefit of any retained air in the pipeline. Consequently a transient pressure rise or shock of potentially damaging proportions can be generated in a pipeline system, even at normal filling rates.

In addition to venting through the Large Orifice (1) when water approach velocities are sub-critical, the Vent-O-Mat series RBX air release valves feature an automatic Anti Shock Orifice (8) device that serves to decelerate water approaching at excessive speed, thereby limiting pressure rise to a maximum of 2× rated working pressure of the valve.

2. Surge alleviation – pipeline pressurized

In instances where a pipeline experiences water column separation due to pump stoppage, high shock pressures can be generated when the separated water column rejoins.

The Vent-O-Mat series RBX takes in air through the unobstructed large orifice when water column separation occurs, but controls the discharge of air through the 'Anti Shock' Orifice as the separated column commences to rejoin. The rejoining impact velocity is thereby sufficiently reduced to prevent an unacceptably high surge pressure in the system. In the same way the series RBX valve prevents high surge pressures resulting from liquid oscillation in a pipeline.

3. Pressurized air release from a full pipeline

Effective discharge by the valve of pressurized air depends on the existence of a 'CRITICAL RELATIONSHIP' between the area of the Small Orifice (7) and the mass of Control Float (4), i.e. the mass of the float must be greater than the force created by the working pressure acting on the orifice area. If the flot is relatively too light or the orifice area relatively too great, the float will be held against the orifice, even when not buoyed, and air discharge will not be effected.

To ensure that the correct 'CRITICAL RELATIONSHIP' exists the requisite 'DROP TEST' must be applied to any air release valve which is intended for discharge of pressurized air.

Venting of a filling pipeline (sub-critical water approach velocity)

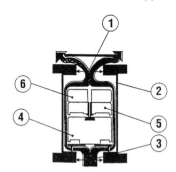

Air enters Orifice (3), travels through the annular space between the cylindrical floats (4), (5) and (6) and the valve Chamber Barrel (2), and discharges from the Large Orifice (1) into atmosphere.

Fig. 29 (above and facing). Operation of Controlled Air Transfer Technology air valve (courtesy Mulric Hydro Projects, South Africa)

Venting of a filling pipeline (excessive water approach velocity)

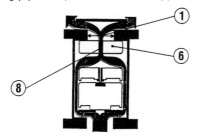

In reaction to increased air flow, Float (6) closes Large Orifice (1) and air is forced through the Anti shock Orifice (8) resulting in deceleration of the approach water due to the resistance of rising air pressure in the valve.

Attention is draw to pre-notes 1 and 2 opposite.

Pressurized air release from a full pipeline

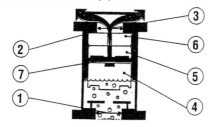

Subsequent to the filling of a pipeline, liquid enters the valve Barrel Chamber (2) and the Floats (4), (5) and (6) are buoyed so that the Large Orifice (1) is closed by Float (6), the valve will then become internally pressurized. A minimal working pressure of <0·5 bar (7·3 psi) acting on the relatively large area of the Orifice (1) will lock Float (6) into the closed position across the Large Orifice (1).

Disentrained air rises through the liquid and accumulates in the valve chamber. When the volume of air is sufficient to displace the liquid, Float (4) will no longer be buoyant and will gravitate downwards, thereby opening the Small Orifice (7) and allowing accumulated air to be discharged into atmosphere. As air is discharged the liquid raises Float (4) to reseal the small Orifice (7) and prevent escape of liquid.

Specific attention is drawn to pre-note 3 opposite.

Vacuum relief (air intake) of a draining pipeline

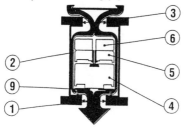

Simultaneous drainage of liquid from Valve Chamber (2) causes Floats (4), (5) and (6) to gravitate downwards onto the Baffle Plate (9), thereby allowing atmospheric air through the valve to rapidly displace draining liquid in the pipeline and prevent potentially damaging internal negative pressure.

Fig. 29 – continued

Controlled Air Transfer Technology (CATT)[35] 'air valves' combine all the above functions of double air valves in one unit. CATT 'air valves' are also designed to prevent surging and water hammer in addition to the air release/entry function (Figs 29–31). A description of the

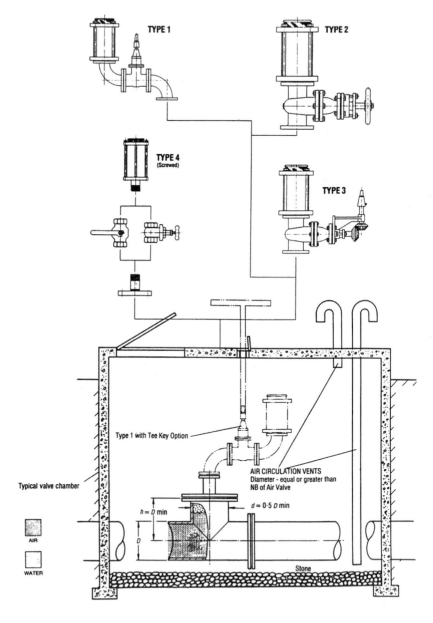

Fig. 30. Typical air valve arrangement

operation of CATT air valves is given in Fig. 29, courtesy of Mulric Hydro Projects, South Africa.

Air valves are essential at the high point of rising branchlines (refer 5.4.1). Air valves must be fitted in the vertical position above an isolating valve. A bevel geared isolating valve is preferred as it can

Type:
Series RBX - Double Orifice (Small & Large Orifice) with Anti Shock Orifice Mechanism

End Connection:
Flange with screwed studs.

Nominal Sizes:
DN80 (3")
DN100 (4")
DN150 (6")
DN200 (8")

Model Nos:
RBX 1601 & 1631 _____
RBX 2501 & 2531 _____
RBX 4001 & 4031 _____

Pressure Ratings:
PN16 (232 psi) ANSI #125
PN25 (363 psi) ANSI #250
PN40 (580 psi) ANSI #300

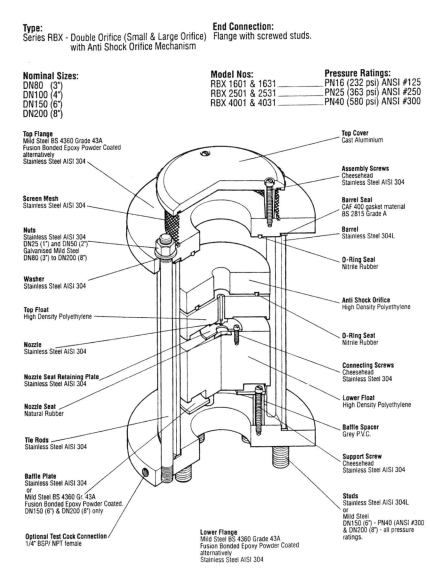

Top Flange
Mild Steel BS 4360 Grade 43A
Fusion Bonded Epoxy Powder Coated
alternatively
Stainless Steel AISI 304

Screen Mesh
Stainless Steel AISI 304

Nuts
Stainless Steel AISI 304
DN25 (1") and DN50 (2")
Galvanised Mild Steel
DN80 (3") to DN200 (8")

Washer
Stainless Steel AISI 304

Top Float
High Density Polyethylene

Nozzle
Stainless Steel AISI 304

Nozzle Seat Retaining Plate
Stainless Steel AISI 304

Nozzle Seat
Natural Rubber

Tie Rods
Stainless Steel AISI 304

Baffle Plate
Stainless Steel AISI 304
or
Mild Steel BS 4360 Gr. 43A
Fusion Bonded Epoxy Powder Coated.
DN150 (6") & DN200 (8") only

Optional Test Cock Connection
1/4" BSP/ NPT female

Lower Flange
Mild Steel BS 4360 Grade 43A
Fusion Bonded Epoxy Powder Coated
alternatively
Stainless Steel AISI 304

Top Cover
Cast Aluminium

Assembly Screws
Cheesehead
Stainless Steel AISI 304

Barrel Seal
CAF 400 gasket material
BS 2815 Grade A

Barrel
Stainless Steel 304L

O-Ring Seal
Nitrile Rubber

Anti Shock Orifice
High Density Polyethylene

O-Ring Seat
Nitrile Rubber

Connecting Screws
Cheesehead
Stainless Steel 304

Lower Float
High Density Polyethylene

Baffle Spacer
Grey P.V.C.

Support Screw
Cheesehead
Stainless Steel AISI 304

Studs
Stainless Steel AISI 304L
or
Mild Steel
DN150 (6") - PN40 (ANSI #300
& DN200 (8") - all pressure
ratings.

Fig. 31. Component description and material specification

be operated outside the chamber, which is vital if the chamber floods as a result of an air valve failure. Double air valves (DAV) are available with an integral stop valve, located between the two air valve bodies. The air flow capacity of an integral stop valve DAV is less than the standard DAV. The integral stop valve DAV also contravenes the principle of discrete isolation valves on all branches on a pipeline. All air valves have a ball or float which floats in the air stream, rising to shut when the air has all escaped and falling to allow air to enter. Some modern designs use levers to control the ball float movement and the floats are made of stainless steel. Valve trim on the modern designs are also stainless steel instead of copper alloy.

Air valves can cause surges by sudden closure/opening and this may result in catastrophic failure of cast iron mains.[36] Regular maintenance of air valves is therefore important. Replacement ball floats are available for most sizes and rubber gaskets for the large orifice valve can be fabricated by a good rubber company. Air valves *must* be installed in chambers to allow them to breathe, to allow expelled water to drain, and to facilitate maintenance. The arrangement of some air valves with low cover on the main may not allow removal of the float balls. This problem can be solved by fitting an air valve with an integral stop valve (e.g. DAV) which needs no separate isolating valve. The DAV sits lower in its chamber giving more room for maintenance access. An alternative solution is to install a lugged wafer butterfly valve for the air valve isolation.

Some modern highway access covers on air valve chamber lids can seal so well that the air can blow a hole in the road surface and even lift it when charging. A superior arrangement is to install a vent pipe from the air valve, terminating above grade with a downward facing elbow (gooseneck). This allows free flow of air through air valves.

Some small single air valves have a large heavy body with a 25 mm or 50 mm threaded connection. These valves require good supports to avoid stressing the small diameter inlet pipework.

5.4.1. Vacuum valves

The function of vacuum valves is to prevent the collapse of pipelines caused by a break at a low point resulting in high outflows, thus creating a partial vacuum. They should be considered for high points on large diameter pipelines which fall steeply, e.g. 100 m fall in 1200 m length.

Vacuum valves for rising mains should be fitted *before* the control valve arrangement at the local reservoir. Rising mains are branchlines which rise up to the local reservoirs. Draining of these lines without an air valve at the high point will result in a partial vacuum in the pipe. Calculations can be made to determine whether or not the pipe will collapse (buckle). However, an air valve will facilitate easy discharge (scouring) and charging of the rising main, which is good waterworks practice.

5.5. Scour valves (drain/blow off valves)

Scour valves need to work when they are needed. They are vital to the emergency operation of a pipeline, e.g. shutdowns. Another requirement is tight shut off. The soft seated resilient sluice valves meet this requirement. Sluice (gate) valves often need a gearbox to allow easy opening/closing against the large forces of unbalanced pressure. Other valve types such as plug or ball function well and need a gearbox to slow down the actuation. This is because the smaller sizes are actuated from shut to open in a quarter of a turn. This would be fine for a pump station but not for buried service.

Scour valves are somewhat misnamed, as a small valve fitted to a large main will not, *in the main*, cause a high enough scour velocity. For example, one 200 mm valve on a 900 mm diameter main might produce only a velocity of 0·7 m/s in the main pipe. True scouring would require large branch valves.

The velocity attained through a small (150 mm or 200 mm) scour valve is very high, approximately 15 m/s if the valve is fully opened. This velocity exceeds the maximum recommended, e.g. 6 m/s for concrete linings. It follows that long-term use of a scour system may result in erosion of the lining and subsequent corrosion of the inlet branch and tail piping. Controlling the flow and hence velocity of a scour is problematical.

The problem is twofold: first, the time for scouring must be a minimum; this determines the rate of flow required; second, the head (pressure) at the scour inlet is often relatively high. Thus the total hydraulic energy of the water to be scoured is a fixed quantity and often high.

Controlling the hydraulic energy or, more correctly, the rate of hydraulic energy is possible but increases the scouring discharge time and so conflicts with the operational requirement for quick times during shutdowns. Also, to achieve this by throttling a scour valve, for instance, has negative effects. Destroying energy in a scour system can be done two ways: throttling the scour valve which is on the inlet pipe connected to the main; installation of a special energy-destroying valve at the exit, such as might be used on a large power scheme dam. These are costly valves and thus are unlikely to be used.

Throttling a scour valve (e.g. sluice valve) causes the gate to rattle and also causes wear and tear on the seat and gate. This is caused by cavitation which is an interesting hydraulic phenomenon of water vapour bubbles forming and collapsing, thus causing pitting erosion.[37] Resilient gates or plugs provide some resistance to cavitation. Cavitation is usually associated with pump operations at low suction heads. Plug valves, while designed for throttling and flow control at normal pressure differentials, may also suffer from the effects of cavitation in the valve body.

A second option is to destroy the hydraulic energy *after* the water has exited the scour pipe. In some situations, this may not be necessary as the water forms a spray which dissipates some of the energy. More likely, the wingwall (refer 5.5.3.1) will be used.

It should be noted that the pipeline resistance of an unthrottled scour system will be low so that cavitation may still occur in the scour pipe system.

Internal inspection of the pipeline and scour branch will reveal any erosion/corrosion problems. One solution could be to use an erosion-resistant pipe material, e.g. stainless steel on the scour valve system.

Scour tail pipes are best constructed of pressure piping, e.g. steel. This allows high velocity discharge which for stormwater grade concrete pipe would exceed its pressure rating. Scour valve arrangements where the tail pipe enters a concrete riser and the exit pipe is at a higher level should be avoided, as the flow is limited to the corresponding head required to lift the access lid on the chamber. On a large main, such a scour system has to be throttled and is very slow to discharge.

Scour valves should be installed in chambers for easy maintenance/ replacement. The piping and the valve should be arranged to allow removal of the valve to facilitate unblocking of the inlet pipe, *when depressurised*, should it be necessary. Plug valves are best fitted with the valve seat upstream, which avoids the build up of detritus in the valve body. Rodding of scour valves under pressure is dangerous and such a procedure prohibited as debris can fly through the valve at high speed when freed.

5.5.1. Bottom scours

Bottom scours are connected to the bottom quadrant of the pipeline (refer Fig. 32) with a valve in the horizontal position. Some scour valve arrangements connect to the very bottom of the pipeline with a bend, but the scour valve is at a lower elevation and requires a

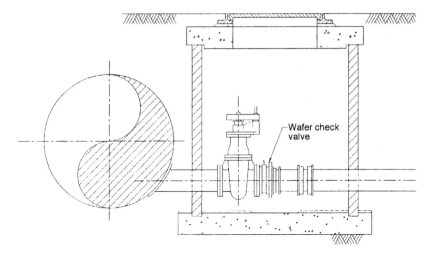

Fig. 32. Bottom scour valve arrangement (courtesy Wellington Regional Council, New Zealand)

deeper chamber. A check valve (typically wafer type) is installed downstream to prevent backflow into the main. The bottom scour free drains to stormwater systems, land or streams.

5.5.2. Top scours

It is not always possible to install a free draining scour system at every low point on the pipeline. For example, the grade of the pipeline may be below the local stormwater system or the pipeline may cross under a river. Top scours should be fitted with a pump-out connection (refer Fig. 33), and the main fitted with a local sump to allow pumping out to a dry condition.

Re-engineering of existing scours that are not completely free-draining can incorporate a top scour with the existing bottom scour feeding a pump-out chamber. In this idea, the top scour is opened first and when the scour runs dry, the bottom scour is opened and the sump is pumped out.

5.5.3. Erosion damage caused by scouring

Scour valves with tail pipes exiting into small creeks or on to dry land may cause damage if opened fully. Erosion of stream banks can be avoided in some instances by installing bends to 'fire' the water parallel to the stream flow. This is not always successful if the stream is narrow and twisty. In this case, an energy dissipater is needed.

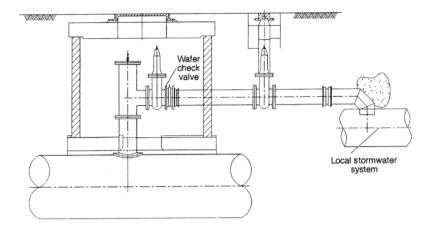

Fig. 33. Top scour valve arrangement (courtesy Wellington Regional Council, New Zealand)

5.5.3.1. Wingwall

This consists of a concrete wall, 'fired' at by the scour exit pipe, thus dissipating the energy in the water. The water falls to the floor at the base of the wingwall and drains away slowly by way of a large outlet pipe.

5.5.4. Dechlorination

The reader's local laws may allow the discharge of potable water to streams and watercourses *provided* that the free or combined chlorine level is zero. Past dechlorination procedures may have consisted of placing bags of sodium thiosulphate in the stream to form a dam and controlling the scour flow so that the water overflowing the dam had zero or 0·1 ppm chlorine residual (ascertained by a colour comparator). This procedure does not work, especially for scours discharging 500 l/s or more. The need to discharge water quickly *and* to meet the requirements of the law leads naturally to the development of real time dechlorination.

The schematic for this operation is detailed in Fig. 34. This is similar to the mobile dechlorination unit designed by Kentucky-American Water Company.[38] The dechlorination equipment consists of a 20 litre sodium thiosulphate solution tank and a 12 V battery-powered dosing pump. The dosing pump discharge pipe connects to the 15 mm threaded connection on the scour valve tail pipe.

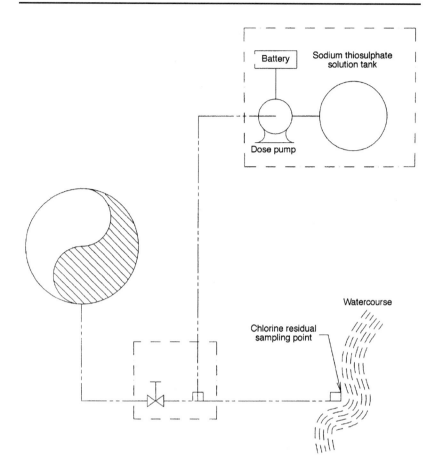

Fig. 34. Dechlorination schematic (courtesy Wellington Regional Council, New Zealand)

The chlorine residual is measured at the scour tail pipe exit, using a chlorine colour comparator. The installation of a pitot tube and sample valve at the end of the scour tail pipe is required to facilitate easy sampling of the exit water. There is a minimum length of scour tail pipe to allow completion of the dechlorination chemical reaction before the water exits into the watercourse. At low velocities, e.g. 2 m/s, this distance is approximately 3 m. At high velocities, when the scour valve is fully open, the minimum distance may exceed the installed distance of the scour tail pipe. The effective distance can be increased by adding a hose to the end of the tail pipe. Alternatively, a static mixing device could be installed just downstream of the dosing point. As the scour velocity decreases over time, the pump, if unadjusted, will overdose the water and begin to give an excess of sodium thiosulphate. Hence, residual chlorine measurements should be carried out every 5 minutes or so.

An automatic control system on the dechlorination equipment was considered by the author but rejected for several reasons. Firstly, the pressure loss in a short scour pipe is so small that the pressure is insufficient to drive (unpumped) a chlorine residual meter. This is usually the case because the meter is mounted higher than the scour. Secondly, chlorine residual meters are sensitive and not considered to be a portable rugged unit for the field. Thirdly, there is added expense and complexity.

Scour tail pipes can be retrofitted with 15 mm threaded connections for chemical injection. Scours which exit into stormwater pipes (which shortly discharge to a watercourse) can also be rebuilt with hard piping (e.g. steel) fitted with the 15 mm injection connections. Similarly, scours discharging on to land can be fitted with tail pipes having dechlorinating connections, allowing the same dechlorination equipment to be used.

5.5.4.1. Dechlorinating new mains
The equipment described in 5.5.4 can also be used to dechlorinate the water in new mains, superchlorinated after construction/pressure testing. The dewatering flow rate has to be low to match the dosing pump's ability to neutralise the chlorine.

5.5.5. Dewatering
Dewatering here refers to the pumped dewatering necessary on a flat graded pipeline, not to gravity scouring. Pipelines with top scours present the biggest challenge, as most of the pipeline will still be full after top scouring. A calculation on the amount of water to be pumped out may for example, give 3 ML between two line valves. This requires serious pumping.

A diesel-powered pumpset (e.g. 150 mm pump), mounted on a trailer will pump at a good dewatering rate, say 0·25 ML/hour. Even such a pump will take 12 hours to pump out, so a couple of pumpsets will be needed. System operational requirements will dictate the maximum time available for inspection/repair. The dechlorination of pump-out water should be done by adding the equipment of 5.5.4 to the pumpsets.

A novel method of dewatering low points at pump-out risers uses compressed air to displace the water through a 'dip pipe' or nozzle.

The air valves on the pipeline to be dewatered must be shut and the nozzle fitted after the top scour runs dry. The air inlet valve can be throttled to control the rate of pump-out flow.

Other places requiring pumps are local dips in pipelines. Some of these will require the large pumpset. Normally, a couple of 75 mm petrol-powered centrifugal pumps or electric submersibles will do the job. For a really dry inspection, a small puddle sucker submersible pump is valuable, removing almost all the water from a low point.

The problem of shut main valve(s) leaking and spoiling the pump-out and/or cross connection(s) leaking is best solved by physically blanking all valves. This should be done for safety reasons anyway (refer 5.1). However, a shut down and scour for external only work (e.g. welding) with a leaking valve can prevent successful welding owing to the water present in the pipe. A threaded socket can be installed upstream of the repair, in the bottom of the pipe, to act as a drain. For welding work inside the pipe, a temporary dam can be installed in the pipe to give enough time to complete dry welding work on the bottom part of the pipe.

5.6. Non-return valves

Non-return valves (NRVs) are also known as check valves or reflux valves (Fig. 35). They allow flow in one direction only. They can be found on scour valve outlet pipes, fitted to prevent backflow into the main pipe. It should be noted that the backflow function is not testable, and therefore not guaranteed with a single non-return valve. Top scours (refer 5.5.2) can be used to ensure the scour tail pipe elevation is above the stormwater or stream level. This ensures no backflow is possible into the main pipe. Another application is on a main itself, outside a booster pump station piped in parallel to the main pipe. Here, the NRV stops circular flow around the activated pump station, which would otherwise occur. NRVs are usually fitted to pump discharges, after the discharge isolation valve. An interesting application of NRVs is in reservoir inlet pipes. This prevents the loss of stored water in the reservoir should the inlet main fail (e.g. in an earthquake). The NRV should be installed immediately next to the reservoir inlet valve. Also, a mechanical joint close to, and upstream of, the NRV should be installed as a weak link to prevent damage to the inlet pipework cast into the reservoir.

NRV styles range from the swing type to the nozzle type, from flanged bodies to wafers. The moving element of some NRVs can be damped to prevent surges. A problem encountered with some wafer type NRVs is that of the valve jamming open against the concrete lining of the tail pipe. The solution is to remove a portion of the concrete lining.

Standard coating on all cast components is a bitumen dip on internal and external surfaces. An alternative coating, which is a fusion-bonded, thermoplastic process is available on special request. All other non-ferrous and non-metallic materials are supplied with their natural finish.

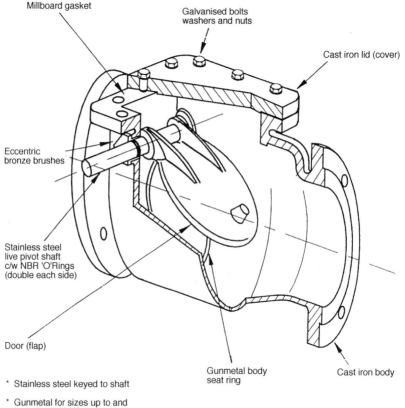

Millboard gasket

Galvanised bolts
washers and nuts

Cast iron lid (cover)

Eccentric
bronze brushes

Stainless steel
live pivot shaft
c/w NBR 'O'Rings
(double each side)

Door (flap)

Gunmetal body
seat ring

Cast iron body

* Stainless steel keyed to shaft

* Gunmetal for sizes up to and
including 100 NB

* Cast iron for all sizes above 100 NB

* Moulded polyurethane resilient facing
for all sizes for positive shut-off at
low differential pressures.

Fig. 35. Non-return valve (courtesy Gillies Foundry & Eng. Co. Ltd, New Zealand)

Maintenance of non-return valves may be neglected. The freedom of the gate to shut/open is important, especially on a pumped line because energy could be wasted.

5.7. Backflow preventers

The function of backflow preventers is to prevent contamination from cross connections to other water systems (Fig. 36). Ideally, bulk water pipelines should never have any such connections. In practice, bulk water pipelines can be connected to private supplies (auxiliary supplies), perhaps as a favour to a landowner when a pipeline was constructed in a rural area, having no town water supply. The selection of backflow preventers depends on the risk from the secondary system. The minimum specification backflow preventer is a double check assembly, typically consisting of a resilient seated inlet valve (ball valve), two check valves (poppet type) with test valves either side and between the checks, and an outlet valve identical to the inlet valve. There is a test valve fitted upstream of the inlet valve to facilitate testing of the check valves by pressurising the downstream of each check valve. The testing of backflow preventers requires special equipment which measures the differential pressure across each valve, in turn. Each check valve must hold 1 psi (7 kPa) differential to pass the test. The volume of water contained in such a system is small, so that a few drops of leakage will cause the differential pressure to fall, thus indicating a failure. The failed check valve must be serviced and retested.

Fig. 36. Backflow preventer (courtesy Febco, USA)

Backflow preventer installations should preferably be above ground, which requires an extra small air valve and a box or cage for protection and for mounting the equipment inside. Backflow preventers with a pressure differential relief valve fitted between the two check valves must be installed above ground.

Testing of backflow preventers has to be by an independently qualified person (IQP). These people may have a plumbing background or a mechanical background and ideally should have trained with a backflow preventer valve manufacturer.

5.8. Emergency shut-off valves

Emergency shut-off valves are fitted primarily in a reservoir outlet pipe in order to preserve the remaining storage should the outlet main fail. They are designed to close on excess flow (velocity) or by sensing seismic movements. Seismic movements do not necessarily directly equate to loss of water whereas the velocity triggered valve does. Some designs are self-contained with a hydraulic accumulator providing the motive force for shut off. Control valves (refer 5.9.4) can also be set up to provide shut off at a set maximum flow. Another design uses a magflo meter to measure the water velocity and the outlet valve is actuated electrically on exceeding the set velocity. Battery back-up power to the electrics is an essential requirement. As in 5.6, the valve should be installed as close as possible to the reservoir, and have a flexible connection immediately downstream. These valves can also be installed on bulk mains to prevent failure of other sections of pipe should a large burst occur. Such bursts might otherwise cause failure under vacuum (refer 5.4.1). A hydraulic analysis of the bulk water pipeline system is required to determine where these shut-off valves go.

Emergency shut-off valve installations are large and must be installed in a separate valve chamber.

5.9. Control valves

Control valves are used to control various flow parameters, e.g. rate of flow; pressure; level (e.g. reservoir level). Several of these functions can be provided by one valve (Fig. 37).

Control valves are either actuated externally or are self-actuating. Actuated valves can be hydraulic, pneumatic or electric powered. Electrically operated valves are common owing to the widespread availability of electric power. Actuated valves require a controller to drive the actuator which achieves the desired output in the parameter(s)

Fig. 37. Pressure reducing valves (courtesy Wellington Regional Council, New Zealand)

set, by modulating the valve position. The valve itself can be any suitable throttling valve, e.g. butterfly, globe, plug, ball etc.

Self-actuating valves require no external power supply. Instead, they use the hydraulic energy inherent in the water pipeline to energise the valve-moving element. It was the piston-actuated valve (which through reinforced rubber technology in the 1950s and 1960s became the diaphragm-actuated control valve) that was designed specifically to answer the requirements of waterworks applications. The roll seal valve is a newer alternative still. The valve consists of a stainless steel cast body and an elastomeric lining which is shaped to roll backwards and forwards. This movement is controlled by the differential pressure between the control chamber and the upstream pressure, in a similar way to the diaphragm-actuated valve. Roll seal valves can be configured similarly to provide various flow functions (refer 5.9.4).

Diaphragm-actuated valves have the underside of the diaphragm directly open to downstream pressure, and only the pressure on top of the diaphragm is controlled by the hydraulic control loop for actuation. These valves, referred to as single chamber, require a minimum pressure differential across the valve in a flowing condition for proper operation of the valve. For on-off control valves this minimum pressure differential is 3·5 m (5 psi), and for regulating valves this minimum pressure differential is 7 m (10 psi). For some applications in which there is very little pressure differential, the internal spring within the main valve assists the valve in the closing cycle. The pressure loss across any given valve is a factor of the valve size at a particular flow rate.

The closing of the valve is performed by control pilots directing upstream pressure into the diaphragm control chamber. The valve opens by the control loop releasing the upstream pressure within the control chamber, to downstream. Speed control of the valve is obtained by controlling the rate of flow into or out of the diaphragm control chamber by way of needle valves or fixed orifices. The opening and/or closing of the valve can be adjusted by means of these speed controls to reduce any possible surging effects to the pipeline system. By locking a pressure within the control chamber that is less than upstream, yet greater than downstream, the valve will position (modulate) somewhere between open and closed. The level of the parameter required (flow and/or pressure) is set by adjusting hydraulic pilot valves on the external control loop. The whole control valve is a dynamic system and can be analysed as such. This system should be a negative feedback system which means the valve will always move in a controlled manner to achieve the output parameter. Then, the valve will stay in the set position until the hydraulic conditions upstream or downstream vary. In contrast, a positive feedback system (loop) is unstable and uncontrollable.

The material in sections 5.9.1 to 5.9.3 is reproduced courtesy of Cla-Val Company, USA.

5.9.1. Rate of flow control valve

The Cla-Val Co. 40-01 Rate of Flow Controller maintains a constant flow rate regardless of changing line pressure. It is a hydraulically operated, pilot-controlled, diaphragm-type globe valve. The pilot control is actuated by the differential pressure produced across an orifice plate installed downstream of the valve. Accurate control is ensured, as very small changes in the controlling differential produce immediate corrective action of the main valve. Rate of flow is adjustable by varying the spring loading on the control. The standard controller includes a calibrated orifice plate and holder that is installed one to five pipe diameters downstream of the valve.

5.9.2. *Pressure reducing valve*

The Cla-Val Co. Pressure Reducing Valve automatically reduces a higher inlet pressure to a steady lower downstream pressure regardless of changing flow rate and/or varying inlet pressure. This valve is an accurate, pilot-operated regulator capable of holding downstream pressure to a predetermined delivery pressure. When downstream pressure exceeds the pressure setting of the control pilot, the main valve and pilot valve close drip tight.

The Pressure Reducing Valve consists of a Cla-Val Co. Hytrol main valve and a pilot control system. The main valve is a single-seated, hydraulically operated, pilot-controlled, diaphragm-type globe valve. The control system is very sensitive to slight pressure changes and immediately controls the main valve to maintain the desired downstream pressure. Pressure setting adjustment is made with a single adjusting screw. The adjustment screw is protected by a screw-type housing, which can be sealed to discourage tampering.

5.9.3. *Altitude valve*

The Cla-Val Co. 210 Altitude Valves control the high water level in storage reservoirs without the need for floats or other devices. They are non-throttling type valves and remain full open until 'shut-off' point in the reservoir is reached. They are constructed to allow easy installation and can be readily protected against adverse weather conditions. There are four basic models available and they are ideally suited for service on reservoirs of any height.

The Cla-Val Co. 210 is an hydraulically operated, pilot-controlled, diaphragm-actuated, modified globe-type valve. The main valve has no packing glands, is single-seated, and has a resilient disc for tight closure. The pilot control operates on the differential in force between the water in the reservoir and an adjustable spring load. This control accurately and precisely closes the valve at the desired water level. The pilot control utilises a diaphragm-actuated three-way valve that alternately supplies pressure to the main valve diaphragm chamber to close the valve or exhaust pressure to allow the valve to open. The pilot control senses the reservoir head by means of a sensing line connected between the pilot control and the reservoir.

5.9.4. *Other set-up variations*

Control valves can be set up with a remote on/off, by installing an electric solenoid valve in the control piping. Once on, the control valve is self-controlling. The solenoid valves can themselves be controlled by time clocks so that the control valve switches on and off on the time clock. A more sophisticated set-up can be achieved by interfacing the control loop with a telemetry system. Should the telemetry fail, the control valve will resume its own hydraulic pilot control.

An adjustable speed non-return function on the control valve can be achieved by installing a non-return valve in the control piping. Control valves can be configured to provide pressure relief on a pipeline system. Control valves may also provide pressure sustaining of the upstream condition, although such a set-up would more likely be used on city reticulation rather than on a bulk water system.

Control valves can be configured as an emergency valve (burst valve, refer 5.8) to shut at a set value of flow. Two directional flow is also possible.

5.9.5. Operating problems

From a pipeline perspective, the critical parameter for control valves is the rate of opening/closing. Too high a rate can cause surges which may burst brittle mains.

Correct sizing of control valves is also important. Oversizing creates a difficulty under low flow conditions because the valve is almost on the seat and can hunt around the set point. One manufacturer offers a V-port insert which extends the life of the seat at low flow conditions. However, the V-port insert results in a greater pressure loss across the valve. Cavitation is another possibility, being a function of the inlet and outlet pressure.[39]

Control valves can cause an overflow condition, caused by small stones and debris jamming the valve partly open. The solution is to install a strainer immediately upstream of the valve and periodically to inspect and maintain it. Similarly, the control loop filters need yearly maintenance.

5.9.6. Installation

Control valves are best installed in a concrete rectangular valve chamber, complete with isolating valves either side, to allow in-situ maintenance or removal. The ancillary equipment is installed in the same chamber. The isolating valves can be installed outside the chamber to minimise the chamber size, but these isolation valves will be subject to buried service duty. Chambers should be fitted with drainage adequate to cope with weepage, seepage and any likely pipe flow into the chamber when the valve is opened/dismantled for service or repair. This avoids flooding with the attendant risk of damage to the ancillary equipment.

The bypass pipe and valve are usually located outside the control valve chamber to optimise the chamber size, which needs to be at least 2·5 m wide for easy access to the equipment. Aluminium cover lids are easier to lift up than galvanised steel ones.

The bypass valve can sometimes also be used to feed water in the reverse direction around the control valve. This mode of operation may be useful for a large reservoir (small consumption) at high elevation which can backfeed another lower elevation reservoir, if the bulk main is shut in an emergency.

6. Cathodic protection

A slightly different approach is taken here and a specification for the provision of cathodic protection (CP) on water pipelines follows. It should be noted that there is little mention of the electrical side of CP. This is deliberate, as the majority of the work is mechanical in achieving continuity and discontinuity. Consultants in CP will provide the detailed design of the electrical equipment.

6.1. PART A: Background to the principles of cathodic protection for water pipelines

6.1.1. Introduction

The purpose of this specification is to set out the requirements for making provision for CP on metallic water pipelines.[40] The CP may be installed at the outset or retrofitted later. It is important that the specification is used at the pipeline design stage once the prior decision has been made that CP may well be desirable. Otherwise, the retrofitting cost of the prerequisite provisions of isolation and continuity may prove to be excessive, compared with the cost of the CP itself.

The specification is in two distinct parts, A and B. Part A (refer 6.1) is an introductory background section covering basic corrosion and CP principles, application to water pipelines and details of the provision requirements. Part B (refer 6.2) is the specification proposed to be used in the pipeline design stage.

No attempt is made to specify the CP design itself. This is best left to an experienced CP specialist provider or, at least, referenced in depth to an approved CP instructional text or course materials. This may prove to be a little more extensive than anticipated if the economically correct approach is used at the outset. However, if the CP provider is presented with a pipeline on which the provision for isolation and continuity have been attended to, the application of CP will probably be relatively straightforward.

6.1.2. Coatings and corrosion

A point of view once held by some pipeline design engineers was: *Why be concerned about corrosion when a high quality coating is specified?* The reality is that no coating on a pipeline of any significant length can be made (or end up) perfect, regardless of careful application and sophisticated inspection procedures. A small percentage of defects will always allow the buried environment access to the pipeline to cause corrosion. Unfortunately, corrosion at these small defects will tend to be more intense per unit area than if there were no coating at all. Consequently, the pitting ratio will be much higher. Naturally, this is very significant for a thin wall structure such as a pipeline in corrosive soil and can result in an accelerated time to perforation. There are four categories of corrosion to which pipelines may be subjected and each is outlined below.

6.1.2.1. Normal galvanic corrosion

Normal galvanic corrosion is the general and most common form of corrosion. The structure surface contains large numbers of adjacent anodic and cathodic areas. By definition, for any two selected adjacent points, the most electronegative area will be *anodic* with respect to the more electropositive or *cathodic* area, and there will be a small driving voltage between them. Under buried conditions in the presence of water, galvanic currents will flow through the soil electrolyte from the anodic to the cathodic areas. These galvanic currents are subject to Ohm's Law, and the metal will ionise (corrode) at the anodic areas, i.e. where the galvanic current *leaves* the surface. As well as the anodic/cathodic pair and the electrolyte, a third element, oxygen, acting as a depolariser, is required for this type of corrosion. The three constituents together form a galvanic cell. Clearly, the more conductive the electrolyte, the greater the cell current flow (by Ohm's Law) and hence the greater the corrosion rate. This explains the extreme corrosivity of sea water.

6.1.2.2. Dissimilar metal corrosion (electrolysis)

Accelerated corrosion can occur when two dissimilar metals are in contact with the more electronegative one suffering corrosion. For example:

- steel pipe connected to copper pipe
- steel pipe to cast iron pipe
- cast iron Gibault joint components to mild steel bolts
- steel pipe to copper earthing by way of connection at pump stations etc.

Where possible, such situations should be 'designed' out or the metals should be electrically isolated using insulating couplings, joints or flanges. The much increased corrosion rate over normal galvanic corrosion is due to the much higher driving voltage (DV) between the two dissimilar metals. In comparison, the DVs between the anodic and cathodic areas of the same metal are small.

6.1.2.3. Stray current electrolysis

A typical example of the cause of stray current electrolysis is an electrified DC rail system with rail earth return. Where the pipeline parallels or crosses a railway, it may act as an alternative parallel conductor. This can occur for some distance away from the railway when the pipeline picks up the earth return current at the crossing point and discharges it at some distant point. Significant corrosion can occur where the current leaves the pipeline. Also, for this reason, CP anode beds of the impressed current type (IC) must always be located well clear of major pipelines.

6.1.2.4. Anaerobic corrosion

Anaerobic corrosion is a type of rapid corrosion that is caused when sulphate reducing bacteria produce sulphides which act as the cell depolariser instead of oxygen. The corrosion environment is generally limited to dark anaerobic muds and clays and is usually associated with a strong organic smell, such as hydrogen sulphide.

6.1.3. Cathodic protection
6.1.3.1. What is cathodic protection?

As discussed above, the corrosion process involves galvanic currents flowing out of anodic areas into the electrolyte (soil) and back into the cathodic areas. A somewhat simplified way of explaining CP is as follows. If a sufficient amount of DC electric current can be made to pass from the electrolyte on to the buried or submerged metal surface from an external power source, then the anodic currents flowing from the buried metal can be reversed. Eventually, on every part of the surface, current flows into the metal. The whole structure becomes a large cathode which by definition is corrosion free. The chemical cathodic reaction which occurs as part of this process is known as cathodic polarisation or simply polarisation. A bonus resulting from this process is that the amount of polarisation can be measured with a suitable voltmeter. A measurable minimum voltage can then be established, above which the 100% conversion to a cathode has been achieved, i.e. corrosion ceases.

6.1.3.2. Provision of cathodic protection

The required flow of electric current on to the structure surface from the electrolyte is achieved by making the structure the negative electrode of a power circuit. By connecting the positive side of the DC power source to a positive buried electrode, current will flow to the pipeline. This positive electrode is called the CP anode. There are two types of anode, namely impressed current (IC) and sacrificial anode (SA). The CP method is known as IC or SA accordingly. The IC method employs special metal alloy anodes of low corrosion rate attached to an external controllable DC power source. The SA method makes use of the electronegative properties of special alloys of zinc and magnesium. No external power source is required because of the natural driving voltage existing between the pipeline and the zinc or magnesium alloy anode. The anode is connected directly to the structure by way of a control station.

Both systems achieve the same objective, i.e. current flow on to the pipeline. However, anode for anode, a far greater current output can be achieved from an IC anode than from an SA anode. This is because of the very low natural voltage difference with an SA anode (less than 1 volt), compared with up to 50 volts employed to drive IC anodes. IC systems are used where 'amperes' are required and SA systems are used when only 'milliamperes' are required. Furthermore, sacrificial anodes are consumed at a much higher rate and therefore have a limited life. In practice, the decision regarding which system to use is generally one of engineering economics.

In the case of IC, the usual power source used is the transformer-rectifier and typical anodes employed for pipelines are made of silicon iron alloy, mixed metal oxide or platinised titanium.

6.1.3.3. Measurement of effectiveness of cathodic protection

The minimum polarised state referred to in section 6.1.3.1 is measured by using a copper/copper sulphate reference electrode and a high

resistance voltmeter. In the case of pipelines, the electrode is positioned on the ground over the pipe at the position selected for measurement. In order to make contact with the pipe at these points, test stations can be installed. These consist of wires connected to the pipeline and running up to a suitable accessible terminal box.

6.1.4. Application of cathodic protection to steel water pipelines

6.1.4.1. Decision to apply cathodic protection

The prime objective of CP on pipelines is to act as a complementary second line of defence against corrosion, the coating being the primary line of defence. The result is a 100% insurance against corrosion and any resulting leakage over the life of the pipeline.

The corrosivity of the environment will obviously influence the decision to apply CP. There is clearly more incentive to proceed with CP when the ground is very corrosive than when corrosivity is very mild. In all cases, the consequences of leakage and its cost is the major consideration. In the case of oil and gas pipelines, these consequential costs can be enormous, i.e. ecological contamination in the case of oil and explosion hazard in the case of gas. For this reason, no risk is considered acceptable for these cases and CP is effectively mandatory in all environments. For water pipelines, however, the decision to apply CP is generally a simple one of economics, i.e. the cost of CP compared with the cost of later ongoing leak repairs and limited asset life.

It should be noted that with modern coatings, normal depreciation rates are very low. A CP system not only mitigates against this low rate but reduces it considerably. This is because CP removes the disbonding effect of crevice corrosion at the coating defect boundaries.

6.1.4.2. Soil corrosivity

In the case of normal galvanic corrosion (refer 6.1.2.1) the main indicator of corrosivity is the soil resistivity. Soil resistivity route surveys are normally one of the first things carried out by a CP provider when considering an application for CP.

6.1.4.2.1. Corrosivity assessment from soil resistivity

The following notes and table form a useful practical aid to decision-making.

Soil resistivity contributes to corrosivity on two bases:

(*a*) absolute values
(*b*) relative values or rates of change.

(a) Absolute value assessment
As natural soil corrosion is galvanic and the corrosion rate is proportional to the magnitudes of the galvanic cell currents, the corrosivity can be directly related to soil or electrolyte conductivity, or its inverse, resistivity. A guide has evolved which assists in the practical evaluation of corrosivity, and this is given below.

Soil resistivity, Ohm-cm	Assessed corrosivity
Under 1000	Severely corrosive
1000 to 5000	Severe to moderate
5000 to 10 000	Moderate to mild
Over 10 000	Generally mild

(b) Relative value assessment
When sudden changes in electrolyte resistivity occur of the order of 10:1 ratio, increased corrosive activity of the 'differential electrolyte' category can be expected and will be superimposed on the natural corrosion rate. Such localised situations would not necessarily follow the absolute corrosivity table and may occur in areas in excess of 10 000 Ohm-cm.

Note: The above standard assessments are only a general guideline. No real security can be derived from an apparent mild interpretation when considering a thin wall structure such as a pipe, and the possibility of extremely localised areas differing from the measured trend.

6.1.4.2.2. Further limitations
The corrosivity table is not generally applicable for the corrosion types discussed under 6.1.2.2 to 6.1.2.4. Those environments will always be corrosive to some extent, even when the soil resistivity is high. Anaerobic corrosion, which can be very severe, is anyway generally associated with low resistivity. Cathodic protection may not be fully effective in coping with stray current electrolysis.

6.1.4.3. N.B. If in doubt

If there is any doubt about whether or not CP should be provided, then the 'provision for CP' as set out in Part B should at least be provided initially, if very high retrofit costs are to be avoided later on.

6.1.4.4. Ductile iron pipes

Ductile iron corrodes at a much lower rate than steel and is less subject to pitting. Ductile iron pipe also has a greater wall thickness *per se* so it can be expected to have a greatly extended life over unprotected steel pipe in a low to moderate corrosive environment. However, in a highly corrosive environment, such as high chloride or anaerobic, corrosion to perforation stage is still possible with ductile iron.

Unfortunately, CP cannot be expected to be fully effective on the loose sleeve system typically provided. This is because corrosive groundwater may ingress under the sleeve at defects and result in oxygen depletion or anaerobic corrosion away from the actual defect and not therefore 'reachable' by CP. Therefore, normal high dielectric coatings are preferable if CP is to be applied to ductile iron pipes. Even so, on a ductile iron pipeline with the loose sleeve system in corrosive ground, CP is still better than none at all.

6.1.4.5. Provision for continuity and isolation

These are the two specific and very important requirements covered in Part B of the Specification, and these are discussed in general below.

6.1.4.5.1. Continuity

Cathodic protection will flow only to those sections of the pipeline that are connected metallically to the negative connection point. Continuously welded pipelines fulfil this requirement naturally. However, continuity is not achieved in a mechanically jointed system. The same applies when couplings are installed in welded pipelines for specific purposes such as expansion joints. In such cases, positive provision must be made by using bonds across the joints and bonding in the components of the couplings as set out in Part B.

6.1.4.5.2. Isolation

The now continuous pipeline must be electrically isolated from all other 'foreign' earthed metalwork in order to avoid serious loss of CP current. Obviously, any foreign metalwork accidentally connected into the CP system will receive an unwanted share of the CP current and, in some cases, this 'foreign structure' may absorb the majority of the current, e.g. earth mats.

Isolation is normally carried out at each end of the pipeline and at any installations on route that may result in current drain, e.g.

- branch connections
- valve chambers (concrete reinforcement)
- cased railway crossings
- connection to pumps at pump stations (electrical earthing)
- stream crossings (anchor block reinforcement).

Isolation is achieved by using isolating flange pairs, special isolating couplings or weld-in isolating joints as set out in Part B.

It should be noted that a plain mechanical coupling cannot be used for isolation. Although theoretically isolating when perfectly aligned and concentric with the rubber rings, isolation is unreliable. In the past, attempts to ensure that a mechanical coupling is isolating through the installation of insulation tape under the follower ring have failed.

6.1.4.6. Position of anode bed

With regard to the position of the anode bed, the CP provider will decide this parameter, taking into account any nearby structures. However, in the case of an IC anode bed, three important requirements must be satisfied.

(a) It must be in an area of low resistivity if high voltages or excessive numbers of anodes are to be avoided.

(b) It must be well offset from the pipeline. The CP provider will calculate the required distance but it will normally be between 100 m and 200 m, i.e. outside the pipeline easement.

(c) It must be well clear of 'foreign structures', e.g. gas lines, other water lines, lead sheathed co-axial cables. Offset distances similar to (b) above.

Attempts to consider IC systems in an urban area are thus usually impractical. However, the minimal clearances needed for sacrificial systems do not present a problem.

6.1.4.7. Test stations

Test stations may be installed in order to measure pipeline potentials or to confirm isolation. Isolation test stations make contact with the pipeline and the isolated structure, e.g. pipeline and bore casing at cased rail crossings. These are specified where required in Part B. Interference test stations are also recommended where the water pipeline crosses another 'foreign' pipeline such as a gas line. This requires the co-operation of both authorities.

6.2. PART B: Specification for the provision of cathodic protection for water pipelines

6.2.1. Continuity bonding

An example of continuity bonding can be seen in Fig. 38.

6.2.1.1. Joint bonding

Where pipe joint bonding is specified, two optional methods are available. These are:

(*a*) the cadweld cable bond system
(*b*) the steel strap system.

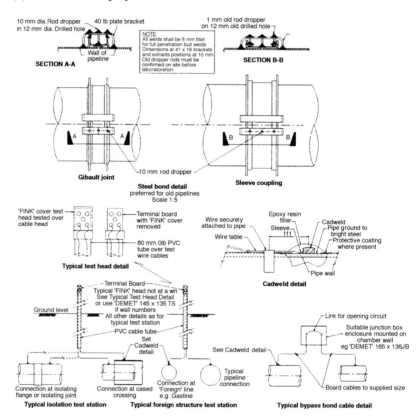

Fig. 38 (above and facing). Cathodic protection details (courtesy Wellington Regional Council, New Zealand)

6.2.1.2. Cable bonding

Generally, type (*a*) is recommended for new pipelines and type (*b*) for retrofitting on older lines. With cable bonding, two jumper bonds should be used across the joint from pipe to pipe. The bonds are ex-PVC insulated copper wire and the bond size will depend on the pipe size and expected CP current drainage. A CP provider should recommend the bond sizing for a major project but, as a general rule, the bond pairs should be not less than $16\,\mathrm{mm}^2$ each on pipes up to 300 mm diameter, and not less than $25\,\mathrm{mm}^2$ on larger pipes. In the case of mechanical couplings, both follower rings and the middle sleeve should be bonded into the system with $16\,\mathrm{mm}^2$ bonds.

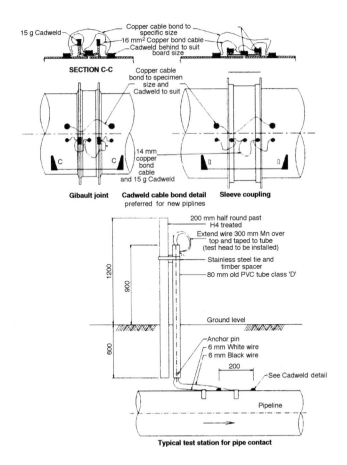

Fig. 38 – *continued*

Preparation of the pipe and installation of the bonds should be strictly to the Cadweld supplier's instructions. In particular, cadwelding cannot be carried out when the steel temperature is below dew point. Heating of the pipe may be necessary in winter. The weld surface must be prepared to clean, bright, steel. The finished weld must be mechanically tested to prove a good connection has been made. The weld must be suitable, encapsulated either with a special proprietary cap or an epoxy type filler. A thorough inspection of cadwelding should be made before the pipeline trench is backfilled.

6.2.1.3. Steel strap bonding

A practical method of steel strap bonding is shown in Fig. 38. The system consists of a suitable steel strap bent at right angles each end to clear the coupling follower rings. Three 13 mm diameter holes are drilled in the strap above the joint follower rings and the middle sleeve. After the strap ends are welded to the pipe each side of the joint, 12 mm diameter steel drop links are dropped through the holes and then welded to the follower rings/middle sleeve and the strap.

The cross-sectional area of the steel strap should be 10 times the recommended copper bond area. In practice, this equates to a 30 mm × 10 mm section minimum for pipes up to 300 mm diameter, and 50 mm × 10 mm section minimum for larger pipes.

6.2.1.4. Bolt bonding

It is important that the coupling bolts are bonded into the CP system. There is normally an automatic mechanical connection to the follower rings, and to assist this, the follower interface area with the galvanised bolt head and nut must have its coating system removed to bright metal for good contact. Washers must not be used. This procedure enables the nut to 'bite in' when tightened. Any exposed steel around the nut and bolt head must then be protected prior to application of the specified coupling protection system.

The most positive method of ensuring bolt bonding is to place a locking nut on the inside of the nut-side follower ring. The locking nut is finally tightened after normal coupling bolt tension is completed.

6.2.1.5. Bypass bonding

Bypass bonding is employed when an isolated section in the pipeline has to be bridged in order to maintain continuity, e.g. at valve chambers, stream crossings or cased railway crossings. The bond should be a suitable size and length of insulated copper cable attached to the pipes on either side by cadweld. The recommended cadweld procedures should be followed (refer 6.2.1.2). The section size of the bond depends largely on its length and the pipe size, and for larger projects a CP provider should advise. As a general rule, the cable size should not be less than 35 mm^2.

6.2.2. Isolation

There are three approved optional methods of achieving isolation. These are flange insulation, isolating couplings and monolithic isolating joints. Each has its advantages in different circumstances.

6.2.2.1. Flange insulation

Flange insulation involves the installation of a proprietary insulating kit between two flanges, a flange and flanged valve or a flange and a flange adaptor (Fig. 39). The kit consists of special plastic sleeves and washers of high dielectric strength, e.g. Mylar, Phenolic, to insulate the bolts from the flanges. The kits are generally available in imperial sizes for use with ANSI flanges. However, they can be readily adapted for use with BS 10 or BS 4504 flanges in most cases. Occasionally, a minor reduction in bolt size may be required. High density neoprene should be used for the insulating gasket. In all cases, insulating flanges should be installed in a chamber in order that checking and maintenance may be carried out.

6.2.2.2. Isolating coupling

The isolating coupling (sleeve type) is very similar to the uninsulated sleeve coupling. It differs by having sealing rings with a skirt (land)

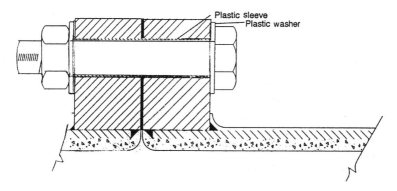

Fig. 39. Flange insulation (courtesy Wellington Regional Council, New Zealand)

on them to prevent the follower ring from contacting the pipe wall. In addition, a rubber stop is provided to keep the pipe ends from touching. Ordinary sleeve couplings *conduct* some electricity because of the conductivity of the sealing rings. Attempts to wrap insulating tape around the outside of the pipe underneath the follower ring do not work, as the sealing ring needs to contact the pipe wall directly in order to seal properly, and this practice does not guarantee isolation.

As with insulating flanges, insulating couplings where possible should be installed in a chamber for testing and maintenance purposes. Where, for economic reasons, couplings must be buried, they shall be protected on the exterior with the petrolatum system of grease and tape. Also, for testing purposes, a surface isolation test station should be installed to measure potentials across the buried coupling.

6.2.2.3. Monolithic isolating joint

The monolithic isolating joint is a factory tested unit which welds directly into the pipeline (Fig. 40). The dielectric medium is a thin layer of resin between the spigot and socket elements of the joint. They are more expensive than a flange pair but they do not require a chamber as no maintenance is possible. The likelihood of isolation failure is extremely low.

Fig. 40. Monolithic isolating joint (courtesy Zunt, Italy)

For confirmation purposes, however, it is desirable to install an isolation test station to measure the potential across the joint. Joints are normally supplied with an epoxy coating ready for direct burial. Field joint wrapping only is required.

6.2.3. Specific provisions
6.2.3.1. Rail crossings
Isolation is required at steel cased crossings under electrified DC railways. Isolation should be inserted in the pipeline a minimum of 10 m each side of the crossing. The isolation is required for two reasons:

(*a*) to minimise the possibility of stray current electrolysis whereby the pipeline can offer a parallel return path for the earth return current with the rails

(*b*) to avoid the very significant problem that would occur if the pipeline crossing accidentally contacted the casing.

Isolation of the rail crossing section can be achieved using the methods detailed in 6.2.2.1 to 6.2.2.3. The monolithic joint is a more expensive option if accessibility or chamber installation is a problem.

Coupling or flange pairs/valves should be installed in easily accessible chambers, typically 1500 mm diameter, to enable maintenance and testing of the electrical isolation. The testing of flange isolation requires access to the whole circumference of both flanges with a minimum of 400 mm between the flanges and the chamber floor. For continuity purposes, a bypass bond should be installed across the isolated section from chamber to chamber to bond the upstream and downstream sections (refer 6.2.1.5). This bond should pass through a suitable junction box in one of the chambers to enable it to be open circuited for later CP section testing. An isolation test station should be installed on both sides of the crossing to measure the potentials across the isolated crossing and casing, to confirm isolation.

In the light of future CP (refer Part A), a special sacrificial anode system to provide CP of the isolated crossing should be considered. Isolation from the casing is mandatory if this is to be effective. The CP provider will advise.

6.2.3.2. Large valve chambers
6.2.3.2.1. New chambers
Typical large valve chambers require the installation of flexible couplings either side of the chamber, to allow for differential settlement.

In order to maintain electrical continuity, the couplings must be bonded across as per 6.2.1.

Positive steps shall be taken to ensure isolation of the pipe from concrete reinforcing steel where the pipe passes through the chamber wall. In the case of electrically actuated valves, the valve must be isolated using flange insulation as per 6.2.2. A bypass bond must be installed across the electrically-actuated valve as per 6.2.1.5.

Where chambers are subject to flooding, all valves must be suitably wrapped for corrosion protection.

6.2.3.2.2. Old chambers

The isolation of the pipe penetrating the chamber walls from the concrete reinforcing steel cannot be guaranteed. The isolating principle is therefore used, requiring the installation of isolating couplings (refer 6.2.2.2) either side of the valve chamber and bypass bonding (refer 6.2.1.5) across the whole of the chamber. It should be noted that the buried pipes between the coupling and the chamber walls will not be protected by the CP system.

6.2.3.3. Stream crossings including bridges

6.2.3.3.1. New crossings

The principle of continuity shall be applied to new stream crossings. Isolation must therefore be achieved between the pipeline and the anchor block reinforcing steel. All mechanical couplings must be bonded into the CP system (refer 6.2.1.1).

In the case of bridge crossings, the pipeline must be isolated from the bridge steelwork pipe supports by the use of separator strips of PVC or high density neoprene.

6.2.3.3.2. Old crossings

The isolation of the pipeline from the existing anchor block reinforcing steel cannot be guaranteed. The isolation principle is therefore used, requiring the installation of isolating couplings (refer 6.2.2.2) before and after the stream crossing. A bypass bond must be installed across the span from just upstream of the upstream coupling to just downstream of the downstream coupling (refer 6.2.1.5). For isolation confirmation, an isolation test station should be installed at each coupling.

6.2.3.4. Tunnels

The pipeline must be made electrically continuous through tunnels and, at the same time, it must be isolated from the pipe supports as for bridges (refer 6.2.3.3).

6.2.3.5. Branch connections

All branch connections should be isolated initially from the main pipe-line for the purpose of having an orderly and controllable CP system. If and when it may be decided to incorporate the branch main into the main CP system, then the branch isolation should be bonded across with a bypass bond. A suitable junction box to provide open circuiting test facilities as per 6.2.3.1 should be provided. Connections not in valve chambers should be isolated at the nearest flange from the main. This is often a buried flanged valve. The bypass valve must also be isolated. Access chambers for maintenance and testing of flange insulation should be provided as per 6.2.3.1.

6.2.3.6. Flanged connections

Non-insulated flange connections will always be conductive to some extent, but may offer some resistance to CP currents owing to the paint at the nut and bolt interface. Where full guaranteed continuity is required across flanges for CP purposes, a bypass bond should be installed across the valve or fitting concerned.

6.2.3.7. Pump stations

Non-isolation at a pump station will result in massive CP current loss owing to the connection to large power authority transformer earthing systems. Where it is decided to include pump station bypass pipework in the CP system rather than insulate at the bypass branches, flange insulation should be installed at the first and last flanges within the station and isolation from reinforcement must be ensured where the pipe enters and leaves the pump station walls.

6.2.3.8. Scour valve and air valve inlet piping

The buried inlet piping for these installations should be protected by the petrolatum tape system to reduce the bare metal the CP system has to protect.

6.2.4. Isolation testing

Testing for isolation should be carried out at least annually and when a fault in the CP system occurs.

6.2.4.1. Local testing

Flange isolation testing where flanges are accessible within chambers or above ground is carried out with a special instrument known as a flange tester (e.g. 'Gas Electronics model 602'). This has two hardened pins which are probed across the flange to register insulation status. The manufacturer's instructions for the instrument must be carefully followed. As well as a straight pass or fail, there is also a procedure for identifying the actual offending short circuiting bolt by 'stepping' around the flanges with the probe.

An important aspect of this testing is the need for good contact between the probe and the flange on both sides (difficult to obtain through rust or a good paint layer). The most reliable way to obtain this is to weld stainless steel 'buttons' on to the steel flanges that are to be tested. Alternatively, a light centre punch mark at each probe position can be made to achieve a bright steel contact point. The tester instrument also has a movable probe for use when testing across insulating couplings and insulating joints.

6.2.4.2. Remote testing

Where the insulation is not accessible directly (e.g. buried), leads from both sides of the flange, coupling or joint are brought to the surface into an isolation test station. Similar testing to section 6.2.4.1 can be done with another type of insulation tester which operates remotely by way of the test wire connections (e.g. 'Gas Electronics model 702'). The instrument manufacturer's instructions should be carefully followed.

6.2.4.3. Isolation testing by potential difference

A further test procedure is available for both local and remote testing utilising a CP reference electrode and corrosion voltmeter. This method can be carried out or advised by the CP provider. It requires some knowledge of CP monitoring methods for proper interpretation.

6.2.4.4. Cathodic protection potential test station

If and when a CP system is installed, proper monitoring requires the ability to get electrical contact with the pipeline at suitable intervals for pipe to soil potential testing. This interval generally varies between 500 m and 1000 m depending on the size and length of the pipeline. With a typical watermain, sufficient contact positions may be available at valve chambers, air valves and scour valves. If this is not the case, then CP test stations should be installed where necessary (refer Fig. 38) for typical test station.

6.2.5. Foreign structure test stations

Where the pipeline directly crosses another major metal structure such as a gas pipeline, steel sewer rising main or lead sheathed co-axial cable, an interference test station should be installed at the crossing point (Fig. 38). The requirements for this station would be a matter for the CP provider to advise. It involves agreed arrangements with the owner of the structure.

6.3. Cathodic protection monitoring

'Monitoring of a CP system is absolutely vital. Even though CP systems are designed to be long term, stable and maintenance free, faults can occur. Cables can be cut through third party interference, lightning can damage a power supply or even destroy the isolation gasket at a flange. Whatever the fault, early detection can most often assist in early repair of a fault. If a cable is damaged, all signs of an excavation may have vanished after two months. . . .

The Natural Gas Corporation of New Zealand Ltd monitoring consists of data collected as follows:

Weekly	Rectifier outputs including voltage, current, variac setting and kilowatt hours
Monthly	CP potentials at key test station locations
Six monthly	Potential readings on all casing pipes and pipelines at cased crossing locations
Five yearly	Close Interval Potential Survey (CIPS) over the pipeline.'[41]

The above monitoring programme can be adapted to water pipelines, bearing in mind that failure, e.g. leaks, have less consequences than in the gas industry.

'By careful monitoring, it is also possible to plot long-term trends (coating degradation and anode consumption being the most common) and make necessary allowances or adjustments as required.

6.4. Coating condition surveys
6.4.1. Close interval potential survey

Close interval potential survey (CIPS) is a technique that is used to record the level of CP being applied to a buried pipeline along the pipeline's entire length. The pipeline potential (voltage) is recorded over every metre of the pipeline using portable computers. Surveys can be conducted using straight ON potentials or, with the CP switching, ON and OFF potentials can be recorded.

To conduct a CIPS, the operator traverses the pipeline route using two probes with an attached $Cu/CuSO_4$ reference cell on each probe which measures the potential difference between the pipeline and the reference. A direct electrical connection to the pipe is maintained by way of a thin copper wire spooling off a specially designed backpack and connected to the pipeline at the nearest test station.

The data once recorded are downloaded directly into a database where a graph of voltage vs distance (metres) is produced. From the graph, coating defects or at risk corrosion areas can be identified where there are significant or prolonged variations in CP potentials. This technique is very effective at locating coating defects or regions on a pipeline where the coating has degraded.

6.4.2. Direct current voltage gradient

Direct current voltage gradient (DCVG) survey is a technique developed in Australia that is used to locate coating defects on buried structures.

An interrupter is installed into the CP circuit causing the current flow to pulse as the interrupter switches at a 300 ms ON, 700 ms OFF cycle. This sets up an oscillating voltage gradient in the soil surrounding a coating defect.

The operator then traverses the pipeline, locating the defect by measuring the voltage gradient between two probes. Accurate location of the defect is carried out by determining the epicentre of the voltage gradient around the defect. The voltage gradient to remote earth can then be measured by adding together the lateral readings away from the defect epicentre until no further voltage gradient is recorded. Defect severity is calculated as a percentage of the voltage gradient measured directly off the pipeline.'[41]

7. Internal inspection

Internal inspection of the lining of pipelines is part of the asset management system which most authorities/companies will have. The time interval between inspections could be 10 years. In some cases, it may not be possible to inspect parts of the pipeline system owing to operational constraints. An analysis of the inspectability of the pipeline system is worth doing to understand the response of the system in an emergency. Then, it may be possible to engineer out some of these constraints.

Inspection without entering a pipeline is possible. Short sections of pipe can be inspected with a video camera. The length able to be inspected is typically 200 m (but recently up to 1 km in the UK[42]) so its use on bulk mains is still very limited. There are three types of propulsion for cameras.

(*a*) Pushing on a cable. This method is difficult to use beyond 100 m and the camera often rotates at bends, which results in upside-down viewing of the pipe interior.
(*b*) Remote operated 'rover' which pulls the camera and its cable into the pipe. These units do not rotate at bends but the length of inspection may be limited by large angle bends.
(*c*) Camera floats in the water (refer to reference 42).

An intelligent independent pig such as the gas industry uses is theoretically possible although storage of video data is needed and the video data would probably be too great for the storage capacity compared with the normal ultrasonic data. These gas pigs are used to gauge the pipe diameter and measure its wall thickness. Water pipes are also lined which probably renders this ultrasonic method void.

The inspection of the metallic pipe wall, even though lined, can be undertaken by a remote field technology intelligent pig but is limited currently to pipes of 150 mm and 200 mm diameter.[43]

In the future, the miniature robots currently being developed by the Japanese could revolutionise internal inspections and repair. 'A high level of autonomous operation will be possible, as will inspections and repairs of pipelines.'[44]

Internal inspection in large diameter pipes is best done personally. The critical perspective is *safety*, because the pipeline constitutes a CONFINED SPACE. 'A permit-required confined space is characterised by one or more of the following elements:

- contains or has the potential to contain a hazardous atmosphere
- contains a material such as grain, sand, or water that has the potential to engulf an entrant
- has a hazardous internal shape such as a sloped flooring that tapers inward and ends in a chute, or similar arrangement. People have been killed when caught in a chute that caused their suffocation by squeezing them in the torso
- contains any other recognised serious safety or health hazard. Included in this last category is a wide range of equipment and dangerous machinery, electrical hazards, or the potential to flood in a heavy downpour.'[45]

7.1. Access points

Access points consist of a short branch 450 mm or 500 mm dia. flanged and blanked, welded on the main pipeline (Fig. 41). The pressure rating of the flanges must be higher than the working pressure of the pipe where they are located. The access point is normally on top of the pipe to allow relatively easy access to the inside of the pipe. Access points on pipes within main valve chambers may well be on the side of the pipe because headroom on a buried valve chamber is too limited. Access to these side entry branches can be made easier by providing a davit to swing the blank away. Otherwise, two people will be needed to struggle with the considerable weight of the lid. Handles are usually welded to the lid to facilitate handling. Gaskets at access points should of course be renewed every time the flanged joint is broken. Air operated impact wrenches provide the quickest and easiest means of breaking/making these flanged joints.

7.1.1. Chambers for access points

Access points should never be buried. If they are, a programme to

Fig. 41. Access point (courtesy Wellington Regional Council, New Zealand)

locate them (from as-built drawings) should be instituted. Chambers should be installed around access points to allow easy access to the flange bolts and, later, to the interior of the pipeline. This is not only good maintenance practice but also saves time in an emergency. Chambers will typically be circular concrete risers with a concrete lid and a 600 mm dia. cast iron entry lid and frame. Chambers should be a minimum of 1500 mm internal diameter.

In common with other features, access points should be located easily using finder drawings and by marker posts in the field.

7.1.2. Location of access points

Access points may have been installed on pipelines with little thought as to their strategic position. An internal inspection will likely be done on a section of main between main valves. It follows that an access point should be provided either side of a main valve or its spectacle blinds. The physical effort of internal inspection on a skateboard determines the distance between access points. A maximum of 400 m is suggested. The other factor is ventilation. It is difficult to get good air without forced ventilation if the access points are too far apart. The second critical place is sumps, e.g. where a pipeline drops underneath an existing stormwater pipe and then rises back up to its former level. The pipe may be fitted with an air valve to bleed air at the 'high point' of the double offset. An access point should be fitted as close as possible to the sump and the air valve should be mounted on the 500 mm dia. blank flange lid. The sump can then be easily pumped out prior to internal inspection. It is possible to fit a pump-out point (say 200 mm dia.) instead of an access point on the top of the main. However, these allow only one pump hose to be fitted and they do not allow visual control of the pumping out of the sump.

A third critical location for access points is at either end of pipeline tunnels. Pipeline tunnels are economic with space and headroom so it is not usually possible to install replacement pipes in the tunnel. There-fore, access is critical to get inside the pipe to carry out welding repairs (steel). Likewise, stream crossings are places to install access points, usually at one side of the crossing, located on the bank, back from one of the headwalls.

It is worth carrying out an analysis of the inspectable pipes *before* any inspection takes place. The cost of additional access points can be estimated and the work scheduled in with other maintenance. The access point can be installed on the main along with its chamber at any time on steel mains, and the final cut-out of the main done later as time/operations permit.

Previously buried access points in roads/highways may not be economic or practical to uncover and chamber up. It may be that an unimpeded location exists close-by, so it is worth installing a new access point. A section of pipeline in the shoulder of a major highway may be considered not inspectable if access is buried or insufficient. However, the pipe either side of such a section will indicate the likely status of the uninspected section of the asset. If the pipes must be inspected, refer to 10.3 for a pipeline modification to create access on these difficult roads.

7.2. Safe system of work

The internal inspection/maintenance of large diameter water pipelines requires a safe system of work, consisting of a Permit procedure (refer 7.2.3), specialised safety equipment and a communications system (Fig. 42). Specialised safety equipment is needed for pipelines smaller in diameter than 2 m or so, to enable retrieval of personnel from inside the pipeline. A motorised skateboard/vehicle is better than a manually powered skateboard or recumbent bikes (quads) which require some effort. Any effort in a confined space as small as a pipe tends to tire people quickly.

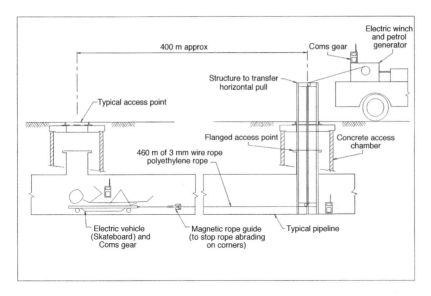

Fig. 42. Pipeline inspection safety equipment schematic (courtesy Wellington Regional Council, New Zealand)

7.2.1. Retrieval system

Two retrieval systems are required, namely a horizontal system and a vertical system. The vertical system is used for lowering/raising personnel in/out of the pipe. The horizontal system is used to retrieve people back to the entry point, if they are unable to reach the next exit point. Vertical retrieval systems are freely available, consisting of a tripod, pulley system and safety harness. Such equipment is typically used for the inspection of sewers.

The horizontal retrieval system advocated by the author consists of a 3 mm, 460 m long rope mounted on a standard winch. The 440 V, 3 ph winch motor is controlled by a variable speed controller. The input power to the controller is a standard 240 V, 1 ph portable generator set. The winch is mounted on the back of a utility vehicle.

The rope guide installed in the entry point after entry by the worker prevents the rope chaffing against the entry point. In addition, magnetic rope guides (for steel and ductile iron pipes) are installed at bends to prevent the retrieval rope from chaffing on the inside of the bend.

The pull exerted by the winch is limited to 100 kg by a 'feasible' weak link of 1·5 mm dia. wire rope between the skateboard and the 3 mm dia. wire rope.

The retrieval system rope is attached securely to the back of the skateboard/vehicle, with the inspector harnessed to the skateboard. Maintenance personnel need to get off the skateboard/vehicle to access the work inside the pipeline. A short, retractable rope must be securely connected to the skateboard/vehicle and the worker's safety harness.

On successful exit at the distant access point, the retrieval rope is disconnected from the skateboard/vehicle and rewound on the winch. The magnetic rope guides installed at bends are collected by the rope and end up on the rope guide at the entry point. The rope guide is then removed from the entry point, along with the magnetic rope guides and the retrieval system moved to the next entry point.

7.2.2. Communication system

There are three types of communication system which each have merits, namely 'Morse code', radio and communication cable.

7.2.2.1. 'Morse code'

The 'Morse code' system uses a series of signals, created, for example, by tapping the side of the pipe with a small hammer. For example, every two minutes, a series of three taps indicates that the worker is comfortable and safe. If the retrieval personnel at the two access points do not receive the signal, then the worker is automatically retrieved. Similarly, a series of six taps means *retrieve me*. With this system, the skateboard/vehicle requires a reverse throttle to allow backing if the worker overshoots an inspection (e.g. a weld). Such a basic communication system is effective and reliable. A small electronic timer carried by the worker inside the pipe is useful to keep time.

7.2.2.2. Radio

Radio communication is an obvious improvement on the 'Morse code' system. The reverse function on the skateboard/vehicle is not required, as the winch operator can inch the worker back to an overshot inspection site (e.g. a weld). This has the advantage of keeping the retrieval rope taut.

Radio equipment for the worker consists of a headset with inbuilt microphone (similar to an airline pilot's), which frees the hands for other duties. The base unit at the access point requires an aerial placed in the main pipeline. The transmitter/receiver probably requires an aerial mounted on the skateboard/vehicle.

The radio equipment operating wavelength (frequency) must be less than the diameter of the pipeline being inspected.

7.2.2.3. 'Talking rope'

'Talking rope' is typically used on intelligent pigs (refer 7), transmitting information back to a base station and allowing retrieval of the pig. For manual internal inspection, the functions of voice communication and retrieval are combined. 'Talking rope' can also transmit low voltage power for lighting and send a video signal back to the base station for recording.

The winch mechanism needs to be larger than the rope only unit and has to be equipped with slip-rings to give rotating electrical contact with the end of the 'talking rope'.

One potential problem is damage while retrieving. Repairs may be difficult and replacement of 460 m of 'talking rope' would be expensive.

7.2.2.4. Combined retrieval rope and communications cable

In this system, the retrieval rope takes the load of personnel being retrieved and the communications cable solely carries communications. The rope and the cable have to be linked together at regular intervals by ties. A large winch drum with slip-rings is required. Any damage to the retrieval rope does not affect the communications cable.

It is possible that this system may not work in practice.

7.2.3. Pipeline Inspection Permit

The following permit procedure is given as a guide and is based on the article 'OSHA Releases New Confined-Space Safety Regs'.[45]

7.2.3.1. Opening of permit

The application to internally inspect a pipeline shall be carried out on the standard form, 'PIPELINE INTERNAL INSPECTION PERMIT' and given to the Pipeline Engineer for approval.

7.2.3.2. Definition

The section of pipeline to be inspected/maintained is to be defined clearly in writing and on drawings, i.e. flow sheets and general arrangements.

- *Define the pipeline to be inspected/maintained.*

7.2.3.3. Purpose of entry

- *Define the purpose, e.g. inspection or maintenance.*

7.2.3.4. Date, time and duration

- *Define the date, time and duration of the work.*

7.2.3.5. Personnel inside the confined space

- *Names of all persons entering the pipeline confined space*
- *Their activities inside the pipeline*
- *How long each will be in the pipeline.*

7.2.3.6. Outside personnel

- *Names of outside personnel*
- *Name of supervisor.*

7.2.3.7. Hazards

- *List all hazards.*

7.2.3.8. Isolation of the pipeline confined-space

The main valves to be shut are to be defined in writing and on drawings. Each of these valves is to be tagged with the name of the senior person who is undertaking the inspection. The tag shows clearly that this person is the only one who can authorise opening or shutting of *any* valves.

Isolation of electrically operated valves

Electrically operated valves shall be fully isolated electrically by the electrical technician. Each valve shall be tagged on the isolated panel. The valve can then be operated manually and tagged by the senior inspecting person.

7.2.3.8.1. Scour valves

All scour valves in the identified section to be inspected shall be opened. These valves shall be identified in writing and on drawings.

7.2.3.8.2. Access points

All the access lids in the identified section to be inspected shall be opened to the free atmosphere. Visible water not drained shall be pumped out from these access points.

7.2.3.8.3. Low points

Low points, e.g. pipe under culverts, shall be identified on drawings and shall be pumped out from the nearest access point.

7.2.3.8.4. Valving procedure
- Main valves are shut, including bypasses.
- Valves to give alternative supply are opened.
- Scour valves are opened.
- Access points are opened up.
- Low points are pumped out.
- Inspection proper begins.

7.2.3.9. Confined space atmosphere tests
- *Details of tests, times and results, testing personnel.*

7.2.3.10. Communication procedures
- *Define communication procedures.*

7.2.3.11. Safety equipment
The following equipment shall be carried and used as a *minimum*:

- safety helmet with inbuilt lamp
- four-element gas detector
- battery powered skateboard/vehicle
- harness and 3 m retractable rope
- communication system
- retrieval system with rope load limiter.

7.2.3.11.1. Retrieval procedure
Any loss in communication with the inspection/maintenance person shall activate retrieval. Any request for retrieval by the person inside the pipe shall activate retrieval. The senior above ground person shall manage all retrievals and shall use discretion to retrieve if they believe the person inside the pipe is having difficulties.

7.2.3.12. Inspection procedure
Inspection or maintenance is carried out by one person only in the pipe *at any one time*. The person in the pipe wears a safety harness permanently connected to the retractable rope (maintenance) or the skateboard/vehicle (inspection only). The retrieval rope is connected to the back of the skateboard/vehicle.

Inspection will begin at an access point on the pipeline at the higher end. It is not advisable to travel 'uphill' against gravity as this is tiring and there is a danger that the skateboard/vehicle will slip inside the pipe.

The inspector/worker will travel from one access point to the next and exit at each open access point for a 'breather'. This also avoids the danger of muscle cramps or claustrophobia. The above ground senior person is to note which section of the pipe the inspector/worker is in and to know this at *all* times.

When the inspector/worker has completed the work and exited the pipe, the senior person whose name is on the valve tags shall authorise the re-valving procedure.

7.2.3.13. Re-valving procedure
- Access points are closed.
- Scour valves are shut.
- Bypass valves are opened to charge the pipe.
- Main line valves are opened.
- Valves to give alternative supply are shut.

7.2.3.14. Notification of inspection completion
The permit holder is to sign and to return the permit to the Pipeline Engineer on completion of the inspection.

8.   Leaks

The repairing of leaks should not be a common occurrence on bulk water pipeline systems built with modern materials. Leaks are of great interest to the Pipeline Engineer as they represent failure, and if investigated will shed light on the status of the asset. The likelihood of future failure can also be estimated. Leaks can be repaired using different methods (refer to chapter 1 under the relevant material, and to 3.2.2.3). They are a challenge and opportunity to the whole team.

8.1. Detecting leaks

Parts of the bulk water system will run in roads/highways and perhaps be confused with city water reticulation by the public. Members of the public who report leaks are often the starting point. One problem is the public's exaggeration of how much water is leaking. They are familiar with domestic flows, not huge bulk water flows. However, it is wise to check out every call even if we think we know what the problem is. This way, the public can see we care about leaks and they will continue to act as an alarm for us.

8.2. Proactive leak detection

Proactive leak detection is theoretically possible on a bulk water pipeline system with sufficient flow meters to compare flows in and out. However, the large flows in a bulk system mean that the water that is unaccounted for might be large, even with high accuracy meters. It is unlikely leaks can be found (deduced) using this method. Even if they can be, their location must still be found, using an acoustic method.

This involves listening for leaks with acoustic transducers placed on the pipeline at set intervals. The distance between transducers can be approximately doubled if the transducer contacts the water. This requires a connection, e.g. hydrant. However, if there is a leak on the detection point, leak detection is not possible until that leak has been repaired. Hence, a leak detection bar is preferred (refer 8.3).

Each leak creates a noise which is picked up by two transducers and correlated between two points of *known distance*. If the distance is not known, the usual procedure is to walk a digital wheel along the route of the pipe from point to point. This can create inaccuracies as the line of the pipe underground is not perfectly knowable. The line of metallic pipes can be located but untraced non-metallic pipes cannot be located. Bends in the pipeline will also create slight inaccuracies in the measured distance and therefore there will be an inaccuracy in the correlated leak position. Other parameters required for leak detection are the pipe diameter and material type. Leak detection equipment can handle composite pipes, e.g. combinations of different materials, sizes and lengths. These parameters must be known.

8.3. Leak detector bars

Leak detector bars consist of a 20 mm or 25 mm diameter round steel bar welded to the top of the pipeline to allow location of the detection transducers. They are fitted as extra features to allow the leak detection equipment to work within its distance limits. They are easier to install than connections, which may require shutdowns on the pipeline. The bars are 'densotaped' for corrosion protection. They finish 100 mm below grade inside a toby box (domestic valve box) and are detailed on plans and indicated by marker posts in the field.

Leak detection bars can be retrofitted at repair sites which have already been dug up, thus saving costs. The repair method can include leak detection bars on non-metallic pipes by using a steel section and two mechanical couplings. The bar can be welded on the steel section in the workshop to save site welding.

8.4. Accuracy of leak detection

Acoustic leak detection is used widely on city reticulation mains and on some bulk water mains which have sufficient features. Correlation of small leaks can be slow and easily missed by a quick operator. A deliberate leak can be created to test the operator. The author created a 14 l/min leak on a 900 mm main carrying approximately 30 ML/day. This was a very small leak in percentage terms. The operator missed it, so a second opportunity to correlate the leak was provided. The leak was eventually correlated and found, even though the correlation took nearly 30 minutes!

8.5. Philosophy of proactive leak detection

The question of the effectiveness of proactive leak detection on bulk mains needs to be asked. 'It is possible that it will prove that leakage from trunk mains is less than originally thought.'[46] A survey initiated by the author of nearly 7 km of 750 mm and 900 mm pipe revealed no leaks on the pipe wall. This is exactly what was expected, as a previous CP survey of soil resistivity showed a high value for the section of pipeline concerned. Also, the leakage history over 40 years was only two leaks, none of them in the section leak detected.

Surveys of pipes protected by CP will indicate future trouble spots (refer to Chapter 6). These ground surveys are thus one step ahead of leak detection and therefore render it largely unnecessary on pipelines protected by CP.

Bulk water mains operate at high pressures, and a leak usually makes its presence known where there are people and roads. The cost of surveys has to be weighed up against the water savings. Here again, the bulk water cost per cubic metre will be relatively low so that the cost of undetected leaks is also low. Leak detection is primarily to save money. On city reticulation systems, it should not be used as an alternative to replacing aged mains that have a high leakage incidence.

8.6. Leakage history

A leakage history should be kept for all pipelines, recording the location and type of leak for each incident. A spreadsheet format is ideal for this. Each pipeline system can have its own sheet.

Such a leakage history allows interpretation of the raw data on leaks. Costs can be shown on the spreadsheet and integrated to show, for example, that a cast iron main should be replaced. The incidence of leakage and failure can be used to control the amount and size/specification of repair materials. For a 110 year old, 750 mm diameter cast iron main, a few lengths of modern pipe and four couplings would be a minimum.

The information on leaks might be brief as described by field staff. Also, there may be no interpretation of the cause offered. Here, visiting the sites of major leaks and writing a formal report which goes on file, is useful.

If your organisation has a computerised asset maintenance system, this can easily produce leakage history reports and other reports on the state of the system. Like all computerised systems, the output is only as good as the data collected and entered.

8.7. Burst detection and leakage control

On large bulk water pipelines, burst pipes can cause major damage to their environs, particularly to roads, and the consequential costs can be high. A burst detection and control system can minimise the damage and the water loss. The monitoring of the pipeline can be achieved by pressure transducers,[47] or by two sensitive flow meters at the beginning and end of the section of pipe under control.[48] In the first method, the fall in pressure below a set value activates the control system to shut down the main valves. In the second method, the two flow rates are compared and the control system actuates the main valves to shut when the differential flow rate is greater than a set value.

8.8. Pressure testing

The purpose of pressure testing a pipeline is to determine that it does not leak at the construction joints, whether welded or mechanical. The straight pipe sections will have been tested in the factory and failures would not be expected when site pressure testing is taking place. The other function of pressure testing is to put a proof stress on the welds. In mild steel welds, this stress will blunt the ends of the microscopic crack-like defects which are part of the weldment.

Civil engineering differs from mechanical engineering in allowing a certain rate of leakage based on the pipe size and length. This procedure seems to have been derived from the hydraulic testing of drain pipes. In mechanical engineering, any discernible fall in pressure is considered to indicate a fault.

Pressure testing of a pipeline requires that the system to be tested should be clearly identified on a schematic drawing. This schematic should show the fill point and the vent point(s) as well as the test pressure, often 1.5 times the design pressure. The schematic can be signed off by the inspecting engineer when the pipeline is tested satisfactorily, to provide part of the construction quality documentation.

Fill points must be the lowest point on the pipeline(s) and the vent points the highest point on the line, in order to ensure the system is full of water with no air pockets before the test pressure is applied. Test ends must be blanked and 'tommed' off to resist the axial force when unrestrained jointed pipes are being tested. In-line valves must be open. Testing against valves is not a good procedure as any leakage will invalidate the test. The valving and any other special pipeline equipment must be capable of taking the full test pressure. If not, these items must be removed from the line and replaced with a straight spool before testing.

New branchlines can be filled often from the nearby bulk main or directly by way of a small bypass valve fitted across the main valve, as part of the original construction.

For small systems of small length, a hand pump will suffice to pressurise the pipeline. Once filled, larger longer pipelines will require a dedicated high pressure, low flow pump. The presence of air in the pipeline system is evidenced by the pulsating of the pressure gauge. Venting can be automatic by using waterworks air valves or manual by using a small gate or ball valve.

Leaks under pressure test can be located with sonic leak detection equipment (refer 8.2) and the value of listening points is then realised. It is not practical to leave trenches open at every joint, so the location of a leak often presents a challenge.

8.8.1. Polyethylene pipe pressure testing

The procedure for pressure testing MDPE plastic pipes is different from metallic and AC materials. PE absorbs energy under pressure (strain energy) and therefore a decay is observed over time from an original pressure. Several measurements and calculations have to be carried out to prove the pressure decays within an acceptable envelope.[49] This creates a conceptual difficulty because a fall in pressure indicates to human senses that there is a problem — not necessarily so here.

9. Modifications to system

Modifications are a challenge to the bulk water team because operations people must work with construction/maintenance people. The critical perspectives are *planning and co-ordination*. The modifications detailed are typical ones. There are many more, one offs, which have deliberately not been included. Problem solving is best done by the people concerned *in their environment*, not by applying an answer from a book of solutions.

9.1. Cut-ins (tie-ins)

Typically, these modifications to the pipeline system consist of new branchline connections, but can be more major in scope, e.g. connecting large diameter inlet and outlet mains from a new terminal reservoir to existing mains. The critical pre-activity is *planning*. The aim of the planning is to make sure the field works are done in the shortest possible time, in order to reduce the risk of water not being supplied to the customers. The design of the pipework can help in this regard. The use of mechanical couplings can shorten the construction time, although anchor blocks will probably be needed. The actual anchor block can sometimes be constructed later, after the cut-in, through the technique of temporarily 'tomming' the affected pipework.

Welded connections are better to make but are much slower than mechanical connections in the larger diameters. Access to the interior of the pipe for welding and lining repair is required and these extra access points add expense to the job.

New branchline connections or scour valve connections can be welded to the main pipeline (steel) without cutting into the main, and an access chamber can be installed around the new valve. Backfilling around the chamber and resealing the road can all be done

Fig. 43. New branches for pump station (courtesy Wellington Regional Council, New Zealand)

beforehand, so that on the day of the cut-in the only activities are cutting into the main, repairing the lining, and reinstating the valve on the inlet pipe branch (Fig. 43).

9.2. Deviations

Pipeline deviations re-route a pipe to allow use of the land in which it is currently sited. For example, pipes may have been laid in open country that is now being subdivided for housing, but the pipes may not lie in the proposed roads. New pipes will have to be laid in the roads and pressure tested and disinfected up to the cut-in points, with the old pipe still in service.

The cut-in points have to be designed to allow construction of the anchor blocks *before* the cut-in is done. This is because the anchor block cannot be constructed on top of the existing pipe, so it has to be located as close as possible to the cut-in point. Each cut-in requires a prefabricated bend (special), preferably flanged at one end to connect to the flange on the deviation.

With two cut-in points per deviation, it makes sense to have two crews working, one at each point, to minimise the time the water supply is interrupted.

9.3. Replacement of main valves

Main valves need to be replaced occasionally to ensure control of the pipeline. The comments on minimising time apply equally here. The arrangement of section 5.1 should be used on pipelines large enough to be inspected.

The new valve and bypass pipework can be prefabricated and assembled offsite so that the installation procedure on the day is as follows.

• Remove old valve.
• Cut off old pipe flange.
• Trial assemble new valve and pipework, tack weld new flange.
• Weld new flange.
• Install new valve, make flange joint and mechanical joint.

The author has organised several of these replacements on 600 mm diameter valves, and the above procedure typically takes 4 hours. The dewatering of the main and recharging of the main inevitably take much longer than this, so the replacement activity is not the major time element.

9.4. Bypasses for control valves

Some reservoirs may have control valves with no bypass valve fitted. It follows that the control valve cannot be removed for major repairs unless the inlet valve to the reservoir is shut. Furthermore, it follows that such a control valve will also not have isolating valves fitted. The usual short distance between the reservoir inlet valve and the control valve presents a difficulty when a bypass connection and a control valve downstream isolating valve are being installed. Sometimes the downstream isolating valve has to be a wafer butterfly type to make this possible. Isolating valves bolted to the control valve need to be anchored by the pipework either side to allow free removal of the control valve. This is not possible in some wafer valve designs. Otherwise, normal sluice valves are used.

The key to smooth installation is prefabrication of as much pipework as possible. Isolation valves for the control valve should preferably be installed in the existing control valve chamber (refer 5.9.6).

9.4.1. Strainers for control valves

Coarse strainers fitted on the upstream of control valves can be useful to catch any stones or pebbles which may have entered the pipeline from the river intake, before the days of full water treatment. Stones can jam the control valve and cause overflowing of the reservoir it controls.

9.5. Installation of magnetic flow meters

Magnetic flow meters (magflos) seem to be a common choice of bulk water meters. They are very accurate over a wide flow range, now 1500:1. One good method of installation on a live system is to shutdown briefly and to cut-in a new length of pipe to suit the magflo diameter. The chamber can then be constructed around the pipe. The final phase is another shutdown and installation of the magflo. This is done by cutting the pipe length previously installed and welding two matching flanges to the magflo. A mechanical joint on the downstream side allows for removal of the magflo.

The internal diameter of the inlet pipe (5D minimum length) and outlet pipe (2D minimum length, both measured from the centre of the magflo), need closely to match the internal diameter of the magflo. The allowable tolerance on a magflo pipe is typically −0 +8 mm. Standard concrete-lined steel pipe seems naturally to match. The only match for non-standard diameters is to hand mortar line a section of rolled steel pipe.

9.6. Installation of sample valves

From time to time, extra sample points are needed on a bulk water main to test and ensure the transmitted water *at that point* meets the drinking water standard set. These small diameter valves are usually installed under pressure. On AC, cast iron, ductile iron and most plastics, the 25 mm diameter connection (typical) is made using a tapping band. The tapping band fits around the main and has a built-in small diameter threaded connection. Tapping machines drill into a main through a valve, first installed on the tapping band. The drill is then withdrawn and the valve closed. The tapping machine is then demounted.

On steel mains, a stainless steel socket is welded on to the main, and the nipple and valve are screwed or welded on to the socket (refer 1.5.2). The connection is made by drilling through the pipe wall of the main, using the valve and socket for internal access. This operation can usually be done under pressure, i.e. on a live main. As above, the drill is then withdrawn through the open valve and the valve is closed.

The handwheels of sample valves should be removed and an operating key given to the laboratory sample collectors. This helps prevent unauthorised opening of sample valves.

A strengthening lug made of steel, welded to the main pipe and the sample valve inlet pipe will help prevent breakage of the inlet pipe caused by people stepping on the valve, and other attempts at vandalism.

10. Tunnels

10.1. Pipelines in tunnels

Pipelines are installed in tunnels for hydraulic reasons, i.e. where the elevation of the tunnel is close to the hydraulic grade line. In other words, there is not enough hydraulic energy to go over the hill so the pipe is installed through it. A useful consequence of this is low gauge pressure in the pipeline so a relatively thin wall pipe is needed (Fig. 44).

The tunnel through which a pipe travels has two ends (portals), although access may be restricted to one portal with the other portal blocked off with a wall and a fresh air vent only. Air flow and quality in such a tunnel may be restricted. Portals may have to be reconstructed to give good access which is certainly required for emergency repairs. It makes sense to improve the access before such an occurrence.

The pipe centreline may be coaxial with the tunnel centreline and this allows access either side of the pipe. If the tunnel dimensions are generous, people can walk almost upright alongside the pipe. Older tunnels may be less generous and people must then stoop and shuffle sideways using a dance step. Such tunnels are much more difficult to work in.

10.1.1. Inspection / Working in tunnels

Periodic inspection of the exterior of the pipeline and the inside of the tunnel requires a safe system of working. Tunnels are a type of confined-space and must be dealt with accordingly. There are two respiratory hazards, low oxygen and gas (e.g. methane). It is possible that both hazards are present together. A four-element gas detector that senses hydrogen sulphide, carbon monoxide, oxygen and methane (inflammables) is recommended. Personnel must take the detector with

Fig. 44. Pipeline in tunnel (courtesy Wellington Regional Council, New Zealand)

them in the tunnel and evacuate if the detector alarm is activated. People should work in pairs and, if in any doubt about air quality, carry a Positive Breathing Apparatus.

10.1.2. Pipe supports

Pipes are typically supported on concrete blocks which support 120° of pipe. Older pipes in tunnels may not be strapped down and could move in a seismic event. Holding down straps should be installed to mitigate against this.

10.1.3. Repair of pipes in tunnels

The repair method depends on the pipe material. Steel pipes may be jointed using mechanical couplings or by means of welded joints. The limited headroom between the top of the pipe and the tunnel invert is likely to be less than the diameter of the pipe. Therefore, replacement of a failed section is not possible unless all the pipe is removed from one portal to the failed section. This is because the pipeline has been constructed one length at a time through the tunnel.

One possibility is to fabricate 1/2 (180°) or 1/3 (120°) pipe shells, about 2 m long, unlined, and construct a new pipe section in situ. The shells can be carried by two people using a skateboard along the top of the pipe. Plates of 5–6 mm thickness will be sufficient even at 1m pipe diameter. An access socket, say 50 mm diameter, can be pre-installed so that welding cables etc. can be passed into the inside of the pipe. After reconstruction of the pipe, hand lining of the interior must be done. Entry to the pipe interior may not be needed but egress is required because the welder will close the pipe. Access points must be provided at either end of a tunnel (refer 7.1.2). Long tunnels with pipes must have regular access points every 400 m minimum for future internal inspection and maintenance, just as a normal buried pipe.

For small diameter pipes (e.g. 400 mm dia.), replacement of a failed section by thin wall stainless steel or plastic (e.g. uPVC, GRP) which is light enough to be lifted manually is a good option. Extra pipe supports may be needed. Repair of leaking mechanical couplings or flanged joints can be achieved using encapsulating repair couplings (refer 3.2.2.3).

10.1.4. Railways in tunnels

Long tunnels, e.g. 3 km and particularly ones that provide the only access to catchment weirs in rugged country may have a small gauge railway installed. These railways not only provide access to the weirs but also allow easy inspection and maintenance of the pipeline, alongside the track. The operation and maintenance of a rail system should mirror that of a larger rail system, e.g. a public railway. The critical perspective is safety and this is ensured by a safe system of work as outlined in chapter 13.

Inspections of locomotives, rolling stock and track should be six monthly. Infrastructure inspections, e.g. tunnels and bridges, should be yearly. Only qualified, i.e. trained, drivers are allowed to operate trains. Records of incidents and accidents are required and will be audited along with other documents every year. Operational standards for the railway will have to be defined and adhered to.

10.2. Tunnels as pipes

Tunnels as pipes will operate on raw water duties only as they are not perfectly closed systems like pipes. Hydraulically, a tunnel may act as a culvert and never be completely full. Concrete lined tunnels will still be porous and the potential for contamination still exists. This is not a problem for raw water duty. Tunnels can operate under gravity or under low lift pump pressure. In the latter case, the tunnel will need to be fully lined to provide a relatively good pressure containing structure.

10.2.1. Inspection / Working in tunnels

Raw water tunnels are a confined space and a similar safe system of work as detailed in 7.2 must be followed. Gas detection equipment must be carried, as it is possible inflammable gas can enter the tunnel when pressures are reduced (as with the Abbeystead pump station, UK). Removal of stones from the floor of tunnels is a major exercise, particularly in small tunnels. Small 'bobcat' diggers will need to be fitted with catalytic converters to purify the diesel exhaust.

Cracks in tunnel walls will need to be grouted using pressure equipment if they are significant. Regular inspection of raw water tunnels close to river intakes is advisable, as stones may have entered the tunnel before the weir intake has been shut off in high water conditions.

10.3. Roads

Working on pipelines in roads presents several challenges to pipeline authorities. Sealed roads are expensive to reinstate and major highways more expensive to reinstate. Budgeting for leak repairs and bursts needs to include a healthy amount for road repairs. In addition, the Highway Authority will stipulate permitted hours of work and these must be complied with. These hours might fall between the morning and afternoon rush hours, e.g. 09.00 until 15.30, but on a busy highway, work might be allowed only after the evening rush hour and until the morning, e.g. 19.00 until 07.00.

In an emergency, such as a large leak or burst, the Highway Authority will be more flexible in their allocation of time, as they have little choice but to allow the prompt repair of the pipe. They may insist the bulk main is shut off and fixed after 19.00. This may be unacceptable if the water system does not have sufficient storage. The smaller repairs or planned maintenance will require thorough planning and logistics in order to complete the work within the 12 hour period. For example, a permit from the Highway Authority must be obtained before work commences. Then, for a major highway, several local radio stations will have to be contacted to broadcast the location of the work and the traffic lanes affected, etc.

Most of the work activities on a critical path analysis will be road works, e.g. sawcutting the asphalt, excavation and backfilling/compaction. These activities are the major cost of the repair. The actual pipeline repair or maintenance is usually the simplest, quickest and cheapest activity.

Logistic reasons dictate that sawcutting of the highway asphalt is carried out the night before the planned work. This may take several hours to complete in addition to the time taken to rearrange traffic flows.

The safe system of work applied to 'working on the road' has a Traffic Management Plan (hereafter TMP) as its main procedure. The TMP will detail the location of the work in relation to the traffic lanes on the road, and show the set-out of the signs and cones to direct traffic. The TMP also details the nominated supervisor who is responsible for the execution of the TMP and the pipeline works. This person must be on site at all times.

On a major highway, the traffic diversion may best be handled by a specialist contractor because large numbers of cones are needed and these contractors are experienced in working on fast, busy roads.

Even small leaks may lead to a large, deep excavation in a road. The original road construction would have employed large rollers to achieve the required compaction. This compaction is not achievable on road repairs using the original basecourse material. A cement stabilised basecourse is needed. Lime stabilised basecourse also works but if moisture is allowed to react with the lime, a cold asphalt top surface tends to break up. On a highway, hot asphalt is required; the cold asphalt is only a temporary surface.

Specialist highway contractors are useful if not mandatory for reinstating the asphalt. They can install friction grip finishes which are approximately 25 mm thick, allowing free drainage of the surface water. These surfaces are likely to be specified for highways. Milling equipment is required to remove the uncut asphalt all around the trench perimeter to allow the 25 mm friction grip finish to key to the original asphalt.

The Highway Authority will want to witness a compaction test and will probably be present for the major phases of the road repair, if not present throughout all the repair. A good relationship with the Highway Authority representatives and engineers is essential.

The Pipeline Engineer must know the standard details of access covers on chambers that are required by the Highway Authority. In highways, these are likely to be the 'gatic' type cover which requires special lifting bars. These covers are bolted to the frame for security. Work on older air valve chambers will require retrofitting these covers to bring them up to the new standard.

The standard fire hydrant cast iron bolted two-piece cover sometimes fails on a major highway. Large trucks put a big load on these covers and their supporting rectangular blocks, if installed proud of the highway surface. These covers can break and cause damage to small air valve installations, e.g. on the 25 mm inlet pipe, and create a large leak. The top support block containing the cover frame tends to crack and disintegrate for the same reason. Repairs at some expense will have to be handled with a TMP. Fabricated steel plate covers are a superior alternative for these small air valve arrangements. Large chambers using 600 mm diameter cast iron lids with frames are also prone to cracking and to the breaking off of small chunks of cast iron. The lids/frames should be inspected yearly, along with the pipeline equipment.

The cost of working on a road may be so high that it is not worth retrofitting a proper chamber around a buried access point (refer 7.1.2). Another possibility is to install a horizontal access pipe (such as a shaft) teed off the main, to a position off the road, and build an access chamber around the blanked end. The access pipe needs to be 750 mm minimum nominal diameter to make access easy. This arrangement means a *once only* dig up of the highway, and future access to the main does not then involve the Highway Authority. These tee sections can be prefabricated to facilitate quick installation.

There would be a negative impact on water quality because these access shafts are dead legs (refer to policy chapter 12).

Likewise, air valves in roads can be reworked to resite them at the side of the road. This is cost effective for long-term maintenance and safety. Double air valve arrangements are often remote, i.e. at the side of the original road. Over time, roads are widened or lanes added so that the air valve ends up in the road again, often sealed over by several layers of road resurfacing. The isolation valve for the air valve is not then accessible, being in the road, and is best relocated off the road for good access. The air valve inlet pipe from the main to the remote position off the road will not then have any isolation at the main. The inlet pipe is effectively part of the main. This arrangement contradicts the philosophy of 5.3, but is necessary for practical reasons. Concern over the mechanical strength of a 100 mm diameter air valve inlet pipe can be allayed partly by using a larger diameter pipe. The concern that the pipe might be damaged by an excavator is somewhat spurious as all pipes can be damaged by careless excavation. Besides, a 100 mm diameter pipe in a major highway is unlikely to be dug up by accident. On a smaller road, the access to the equipment in the road is less of a problem, so that resiting equipment at the road side is probably not worth doing.

10.4. Access tracks

Access tracks are metalled roads, possibly of dubious quality, which give access to a bulk water pipeline. The pipeline might be in the body of the road for some distance. The administration of a track could well be a joint one, say between a farmer and the Pipeline Authority. The uses of the track thus differ. The Pipeline Authority will need a good track to enable trucks and machinery to access the pipe. The farmer may use only four-wheel drive motorbikes and utility vehicles. The costs of track upgrades required for pipeline maintenance will probably be the Pipeline Authority's responsibility.

10.4.1. Scouring of steep sections

Heavy rainfall can scour deep channels in a track and make it difficult for vehicles to pass. Run-off channels and culverts crossing the track are typical minor civil engineering activities in which pipeline authorities get involved.

More serious is the scouring of the cover material over the pipeline, sometimes to the point of exposing the coating. Water then tends to travel down and along the bedding of the pipe trench, leading to undermining. This problem can be addressed by a good sized culvert pipe and intake structure, correctly positioned to collect the run-off water.

Access tracks can get overgrown with grass and weeds if little used. The remedy is to remove the top layer with an excavator and to lay new metal on the exposed sound base.

10.5. Pipeline tracks

Pipeline tracks enable location of a pipeline through hill areas, forests, bush, etc. The main maintenance function is to cut the grass regularly. Steep, overgrown tracks will have to be cleared by hand using scrub cutters and chainsaws for larger trees. Trees growing directly above a pipeline or close by are a problem because the roots grow around the pipe and can potentially damage the coating. A tree allowed to grow on top of a pipe in a populated area will attract considerable attention if left to grow tall. Members of the public might object to its being cut down. Some tree species may even be a protected (by law of the land) variety.

Steep, overgrown tracks can be cleared very effectively using a hydraulic powered mulcher, mounted on a tracked excavator. These mulchers can deal with 150 mm diameter tree trunks and easily deal with scrub, even 4 m high. One advantage of mulching is that there is no vegetation to remove. Once cleared, a track should be regularly mowed, typically with a flail mower mounted on a four-wheel drive tractor. The centreline of the pipe can then be delineated with marker posts to facilitate easy tracking of the pipeline. Really steep pipeline tracks, e.g. 1:2 gradient, will have to be maintained by hand.

The scouring of steep tracks by rainwater is another typical problem. A solution is to install pipe culverts across and under the track, and perhaps an intake structure. These constructions may be outside the present easement (refer chapter 15) so an extension to the easement will be required to enable legal access and give legal protection.

10.6. Bridges

The bridges that support bulk water pipelines may belong to a Highway Authority or they may be a dedicated structure to support the pipe only, thereby being the responsibility of the Pipeline Authority. Pipeline maintenance work on road bridges supporting pipes requires a permit from the Highway Authority and the preparation of a TMP.

Advantage can sometimes be taken of access scaffolding erected for repair/maintenance of a road bridge. The scaffolding may also give access to the bulk water main and allow repainting to take place (for example). The main may also have an air valve that can be adapted to provide a private water supply (refer 5.7) for the civil engineering work on the bridge structure. This 'mutual aid' can save the majority of the cost of scaffolding, which is a major cost in painting maintenance.

Cranes may be needed in addition to access scaffolding or hydraulic work platforms mounted on trucks. This latter equipment is the only practical way to work on pipelines which are slung underneath a road deck, many metres above the ground. The decks are often an inverted W-shape and there is little access to even construct, let alone to maintain the pipe. Pipes may be welded because there is insufficient room to fit a sleeve coupling. The repair clamp of the lip seal type is really useful in such situations (refer 3.2.2.3.3).

Bridges owned by the Pipeline Authority require regular inspection and maintenance, mainly of the steelwork paint, bolts, and pipe supports. Repainting of these structures can be contracted out and will involve access scaffolding.

10.6.1. Stream crossings

Stream crossings typically consist of a length of pipe across a small stream, supported by two headwalls, one either side of the stream. The connection of the crossing pipe to the rest of the pipeline is usually by mechanical couplings, buried next to the headwalls. On cast iron or ductile iron pipelines, the crossing pipe is likely to be steel for strength reasons. On larger diameter crossings where the exposed pipe elevation is higher than the buried pipe elevation, the headwalls are larger and are designed as anchor blocks. These larger diameter crossings are often supported by ring beam fabricated steel pipe supports, bolted down to the anchor blocks. Mechanical couplings are fitted on the stream bank side of the pipe support to allow thermal expansion of the exposed pipe. Another feature of stream crossings which are a low point in the pipeline is scour valves (refer 5.5.1).

Maintenance of stream crossings involves regular inspections, especially after heavy rain which can erode the stream banks and undermine the headwalls/anchor blocks. It may be necessary to strengthen the upstream banks with gabion baskets filled with stones, in order to protect the headwalls. Concrete headwalls can crack and split, and these will eventually need to be repaired by drilling and grouting steel reinforcing bars, or, in the extreme case, by rebuilding the headwall which will also require temporary support of the crossing.

Other regular maintenance work is protection of the exposed pipe coating (refer 2.2.7). An alternative maintenance saving modification is to deviate the pipeline under the stream bed. The pipe can be covered with concrete to protect it. This action also improves the seismic resistance of the pipeline. A top scour valve arrangement (refer 5.5.2) is required. Small stream crossings will need only one length of pipe, thus avoiding the need for joints.

11. Pipe laying/routeing

11.1. Above ground or below ground?

Pipelines can be laid above ground in some cases, although the majority of pipelines will be installed underground 'out of the way'. Table 4 highlights the advantages and disadvantages of each option.

11.2. Pipe routeing

The topography through which a pipe traverses dictates the places where the pipe cover will be deeper than the average. The depth to which a pipe is laid depends on the climate. Colder climates that have deep frosts require deep trenching to avoid the pipes bursting from freezing effects. A minimum cover is required to protect the pipeline from traffic, typically 750 mm to 1 m.

The number of vertical bends and the length of pipe laid deeper than the standard cover are interdependent variables. The solution of a long section (cross-section) will likely be a compromise between the two variables. Plotting a solution to keep the cover the same (say 1 m) will result in the maximum number of bends. Small changes in grade, e.g. short terraces, may not require any bends if a slightly different grade, giving deeper than average cover for a certain length, is used.

Existing services, e.g. gas, power, telecommunications, etc., will require to be surveyed for depth and plan position and the new pipeline route designed to try and naturally clear these obstructions. Where this is not possible, the pipeline will have to deviate from the chosen elevation, below the existing service and rise back up to the same elevation. These double offsets will require air valves either side. They may also be fitted with access points for internal inspection as they constitute a sump, a potential trap for debris.

Table 4. Comparison of above ground against underground pipes

Above ground	Underground
Visible	Invisible
Heat ingress	Little heat ingress
Heat egress	Little heat egress
Pipe supports needed	No pipe supports needed
Thicker wall pipe	Thinner wall pipe
Coating attacked by ultraviolet light	Ultraviolet protected
No soil stress on coating	Soil stress on coating
No cathodic protection possible	Cathodic protection may be required
Subject to weathering, e.g. rain	Subject to groundwaters

Pipes laid need to be free draining so a minimum gradient is required, e.g. 1:500. The long section of a pipeline can be likened to a series of valleys connected together. In the bottom of each valley, a scour valve is required, and on the peaks, an air valve. On a long 'downhill' run, e.g. 5 km, there may be several scour valves in addition to the true low point scour, in order to allow fast draining of the pipeline.

The topography also determines the pressure rating of the pipe and fittings, on a gravity system, which is not necessarily uniform. On high sections, near the hydraulic grade line, the working pressure will be low. On low sections, e.g. close to sea level, the working pressure will be high. The pipeline might constitute a composite with pipes of the same diameter but differing wall thickness (i.e. pressure rating). This is more economic than using a uniform highest pressure rating throughout the whole pipeline.

Generally, the most direct route is preferred for economic reasons. To run pipelines in roads is also common practice because easements are not required and there is consequently less disruption to land-owners. Traffic disruption during construction can be significant though. The routeing of a pipe in a road, preferably in the shoulder, has some advantages. First, the road provides easy access for future maintenance. Second, any leaks penetrating the road surface will be evident and will very likely be reported by members of the public. The disadvantages of pipes in roads are the higher construction costs and the high costs of road reinstatement with attendant traffic disruption, when maintenance is being carried out.

It is logical to install pipelines in the roads of new subdivisions. Pipes that are not in roads may have to be deviated later to allow subdivision to occur, though the expense will likely be borne by the developer. The Pipeline Authority needs to provide input at the planning stage to avoid compromises of the existing easements. Easements on large pipes at the rear of three or four houses in suburbia are definitely not recommended.

11.3. Excavation

Excavations deeper than 1·5 m may be notifiable to the Health and Safety Office of the Government, under Construction Acts or Regulations.[50] The safety of excavations is the responsibility of all workers but is specifically the responsibility of the Safety Supervisor. Trench walls can easily collapse, especially in areas of high level groundwater. Dewatering points must be established. There are three typical methods for preventing trench collapse: wide batter of the trench sides; terracing the trench sides; installation of shoring against the trench sides (e.g. sheet piling and cross-bracing).

Excavation in virgin ground and in areas where there are no underground services is typically achieved using hydraulic track mounted excavators. Rubber tyred excavators (front or back-hoe) are useful for work on roads and can be driven to site. Track mounted excavators must be transported on low loaders which often costs more and requires plates to be put down to protect asphalt surfaces. Excavation of rock requires additional heavy equipment and sometimes the use of explosives, with attendant safety procedures.

The trench width is not much wider than the diameter of pipe to be installed. Pits will be required at joints to give construction clearance when joints are being installed or welded. A typical cross-section is shown in Fig. 45. The grade of a trench must obviously match the pipe grade and can be checked with a laser level. Slight changes in line can often be accommodated by the flexibility of mechanical couplings or joints. Major changes in large angles can easily be overcome on a steel pipeline by fabricating the bend to the correct angle in the field. The disadvantage of fixed angle pipe fittings is that they determine the line of the pipe, to a certain degree. Field fabrication of bends allows the best alignment of the pipe and flexibility in overcoming unforeseen obstructions.

11.4. Bedding and backfill

The purpose of bedding and backfill is to support the pipe evenly and to reinstate the load bearing ability of the area above the pipe, especially if a road is above. The load bearing capacity is produced

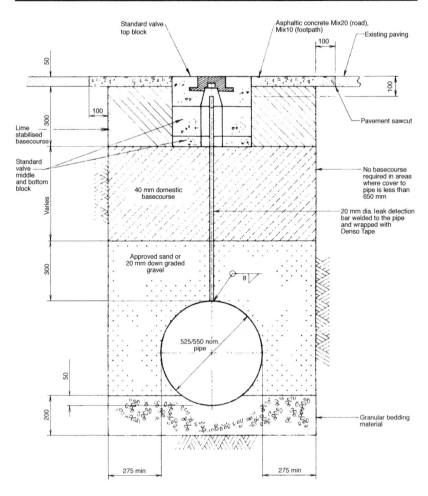

Standard valve top block

Asphaltic concrete Mix20 (road), Mix10 (footpath)

Existing paving

100

50

100

100

Pavement sawcut

Lime stabilised basecourse

300

100

Standard valve middle and bottom block

40 mm domestic basecourse

No basecourse required in areas where cover to pipe is less than 650 mm

Varies

20 mm dia. leak detection bar welded to the pipe and wrapped with Denso Tape

Approved sand or 20 mm down graded gravel

300

8

525/550 nom. pipe

50

200

Granular bedding material

275 min

275 min

Fig. 45. Typical trench section (courtesy Wellington Regional Council, New Zealand)

by compaction, which is usually carried out on several layers, one at a time. Highway authorities will typically require cone penetrometer tests to prove the required standard of compaction (Fig. 45).

The bedding and backfill also act in concert with the pipe to resist vertical loads, especially on flexible pipe design.[51] Sand has been used for pipeline bedding and is still used for plastic pipes because it is a soft uniform material. Typical backfill for metallic pipes is gravel, 20 mm down, which does not wash away or meld with the surrounding trench as sand tends to do.

For pipes not in roads, some of the original excavated material may be reusable as backfill. Large boulders, rocks, concrete slabs, etc. should not be used as backfill because the pipeline coating may be damaged.

12. Water quality issues

Water quality issues can be addressed under ISO 9001 and 9002. The pipeline transmission system is part of the total bulk water system and it impacts on the water quality delivered. For example, the internal lining of the pipe might affect the water quality by virtue of the material of which it is made. Sampling and testing of scrapes from the lining of pipes for coeliforms should be routine whenever a pipe is opened up. It gives an indication of the health of the mains.

The quality certification of a water supply system to the ISO 9000 series of standards consists of several stages:

(a) water treatment plants only, with the focus on product, i.e. treated water as the output
(b) the whole water system, which includes the bulk water pipelines to ISO 9002
(c) the whole system to ISO 9001, where the design of the system must be certified to prove it 'makes' the specified product.

This last stage, full compliance with ISO 9001, involves the design and construction aspects of bulk water pipelines needed to achieve the required water quality.

ISO 9002 requires procedures to be written to control the output (water quality primarily). The proof of the quality system is achieved by internal and external auditing and requires quality records to be kept and maintained. The auditing process allows, and should welcome, improvements. Improvements may not only be to procedure, i.e. working practice, but also to 'design'. For example, the removal of tee sections in a pipeline and replacement with a bend (elbow) to eliminate a previous dead leg (stagnant section of water) is an improvement in procedure and working practice. It is also an improvement in 'design' because such a dead leg might have been avoidable if quality has been part of the original scheme.

It is clear then that the quality paradigm (quality system) cannot be completely successfully applied to a historic pipeline system that probably had an engineering paradigm of a working, functional water system as its objective. Writing procedures for ISO 9002, where there are probably few existing procedures, is always retroactive. To achieve ISO 9001, where design must be proven, is then possible only if the quality design parameters are retroapplied (re-engineered) to the whole pipeline system. It should be noted that re-engineering may be required to meet the water quality anyway whether ISO 9001 is used or not. This would likely be very costly and difficult. An analogous situation would be that of making an older car, say 40 years old, as safe as a modern car and meeting the current legislation regarding new car safety.

Examples are the outlawing of the old ball hydrant, which also acted as an exposed air valve. These would have to be removed and blanked off or replaced with the modern screw down hydrant, and no doubt most have been. Existing air valve installations in roads would have to be relocated to allow the chamber to drain freely and to breathe properly, thereby to avoid pressurising the chamber or drawing foreign material into the main under vacuum conditions. Here, a procedure for air valves cannot be written for ISO 9002 because an air valve is designed to be autonomous, i.e. self-acting, self-regulating. No procedure can therefore be written that forms part of the daily working of the valve, unlike a process in a water treatment plant. Therefore, if the air valve is covered with water (e.g. groundwater), it needs to be redesigned. ISO 9002 does cover minor redesign. These minor redesigns should not be confused with the rigorous design process of ISO 9001.

The structure of ISO 9001 puts the design parameters for quality up front so that the hardware (pipeline system) and the software (procedures) meet the required quality output. It is the author's opinion that this can be totally achieved only on new pipeline schemes.

The following procedures cover the ISO 9002 standard (water quality as the output) as best as the author can currently determine. These are followed by some design problems (parameters) for consideration in achieving ISO 9001 status. The 'standard' ISO 9002 procedure framework has been adopted, i.e.

- objective
- purpose

- responsibility
- actual procedure(s).

12.1. Water freshness control
12.1.1. Parallel mains
12.1.1.1. Objective
- The objective is to keep water fresh in all parallel mains.

12.1.1.2. Purpose
- The purpose is to keep water flowing in all parallel mains.

12.1.1.3. Responsibility
- Senior Supervisor

12.1.1.4. Procedure
The valves listed as open shall be checked *once a year* and reopened where shut. Reference shall be made to Flow Scheme Drawing for overall location and to individual control sheets for exact location.

The procedure applies to the following pipelines:

- *Name the pipeline subsystem.*
- *Name the valves in the subsystem.*
- *All valves shall be open.*

The purpose of the individual control sheets is to function as the quality record(s) for the above procedure. The control sheet has a section for the responsible person to sign. Thus, the procedure can be audited. Fig. 46 details a typical parallel mains quality control record.

12.1.2. Pump station pipework
This is included as pump stations are part of a bulk water system, though they may be under the supervision of another person (e.g. another engineer).

12.1.2.1. Objective
- The objective is to keep water fresh in all parallel pipework.

12.1.2.2. Purpose
- The purpose is to turnover water in all parallel pipework.

12.1.2.3. Responsibility
- Technician

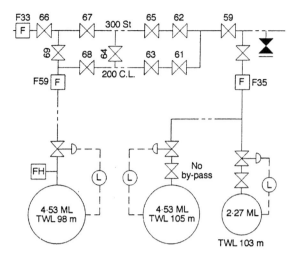

	Valve Location
61	North end Ascot Park, down bank
62	
63	
64	50 m down track below Paramata No 2 Res
65	
67	
68	Top of track at Paramata Res
69	

Certification Nominated valves are open	
Valve Operator	Date

Fig. 46. Parallel mains quality control record (courtesy Wellington Regional Council, New Zealand)

12.1.2.4. Procedure
- This procedure applies to all bulk water pump stations.
- Duty pumps shall be rotated at monthly intervals.

12.1.3. Dead sections in pipelines
12.1.3.1. Objective
- The objective is to ensure stagnant water in dead sections does not mix with fresh treated water.

12.1.3.2. Purpose
- The purpose is to remove stagnant water from dead sections before connecting to live sections.

12.1.3.3. Responsibility

• Senior Supervisor or Supervisor

12.1.3.4. Procedure

The procedure consists of isolating the stagnant section of water (if not already isolated), and using fresh water to displace the old water through a hydrant until only fresh water exits the hydrant.

This procedure is proved on chlorinated water by achieving the same chlorine residual in the water exiting the hydrant as the water upstream of the charging valve. Fig. 47 details a typical dead section quality control record. The procedure is as follows.

1. Open the hydrant between valves 26A and 26B.
2. Crack bypass valves on valves 26A and 26B.
3. Allow the hydrant to discharge for 2 minutes.

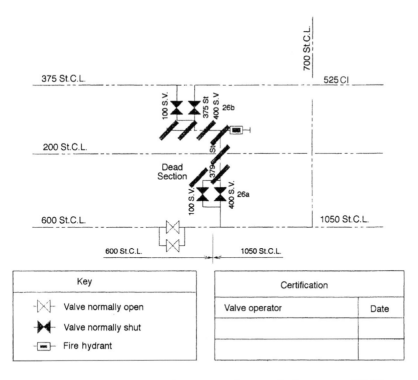

Fig. 47. Dead section quality control record (courtesy Wellington Regional Council, New Zealand)

4. Test the residual chlorine of a sample from outside the dead section, taken from either the 525CI pipeline or the 1050StCL pipeline, depending on flow direction required.
5. Test the residual chlorine in the water discharging from the hydrant.
6. Compare the test results from 4 and 5.
7. If the chlorine residuals are equal in value, valves 26A and 26B can then be opened. If the chlorine residual from the dead section is lower than the chlorine residual from the live section, keep the hydrant discharge going until samples from the hydrant yield the same chlorine residual as the value in 4 above.
8. The hydrant shall then be shut.

12.2. Water contamination control
12.2.1. Private supplies
12.2.1.1. Objective
• The objective is to prevent contamination of bulk water pipes from other connections.

12.2.1.2. Purpose
• The purpose is to prevent backflow into bulk water pipelines.

12.2.1.3. Responsibility
• Pipeline Engineer

12.2.1.4. Procedure
The private supplies listed below shall be tested *once a year* only by an independently qualified person (IQP). Both IQP registrations and test records for each installation are kept on file.

Check valves that fail the test shall be serviced and retested by the IQP. The test records shall detail the two test results for each installation, one for each check valve. Check valves unable to be tested or repaired shall be noted on the record form. The Pipeline Engineer is responsible for deciding whether or not to shut down a private supply, although the emphasis is on repair and retesting the backflow preventer as soon as possible.

• *List all your installations and their locations.*

12.2.2. Excavation for repairs, opening of pipes[52]
Flushing after repair
The AWWA Standard C601-81, Section 9.3, describes the flushing of water mains after repairs to remove any contamination introduced

during repair work.[23] This procedure easily works for city reticulation pipes, where there are hydrants close to block valves.

For smaller bulk mains, e.g. 300 mm dia. mains, with line valves every kilometre, the flushing procedure may also work. However, hydrants on bulk mains are often scarce and the distance between line valves may be up to 5 km. The existing scour valves can obviously be used to discharge contaminated water.

Flushing is possible on larger bulk water mains but it wastes large volumes of water and takes a considerable length of time. This operational time may not be available in addition to the time taken to repair the line.

In conclusion, the author believes that the flushing of bulk water pipelines is problematical and that more dialogue on the subject is needed. The installation of sample tappings and valves at line valves to enable samples to be taken before main valves are open is a companion issue. Any design/practice changes should be incorporated into the ISO 9001 Policy (see 12.3).

12.2.2.1. Objective

- The objective is to prevent contamination of bulk water pipes from the entry of outside matter into the pipelines.

12.2.2.2. Purpose

- The purpose is to prevent groundwater and foreign material from entering bulk water pipelines.

12.2.2.3. Responsibility

- Supervisor or Foreperson

12.2.2.4. Procedure

12.2.2.4.1. Leaks

First, a decision to shut the pipeline down or keep it going must be made. Large bursts must be shut down as soon as possible for safety reasons and to minimise damage. This decision determines the procedure used.

12.2.2.4.1.1. Excavations when pipeline is pressurised. The leak shall continue until a dewatering point is established *below* the bottom of the pipe. Dewatering pump capacity shall be greater than leak flow rate. Scour valves can then be opened.

12.2.2.4.1.2. Excavations when pipeline is shut down. Excavation shall proceed until the pipe becomes visible. High test hypochlorite (calcium hypochlorite) (HTH) powder shall be spread in the trench around the pipe. If the excavation exposes oil or oily material, the HTH powder shall be dissolved in plenty of water before it is spread in the trench.

12.2.2.4.2. Opening pipes

Joints 'cracked' to allow water to exit shall be handled as 12.2.2.4.1. The joint(s) must not be dismantled and a section of pipe removed *unless* there is a dewatering point below the bottom of the pipe and the pumps keep the water level below the bottom of the pipe.

12.2.2.4.3. When the pipe is open

If groundwater seeps into the excavation, HTH powder shall be spread in the trench near the cut ends of the pipe. If the excavation exposes oil or oily material, the HTH powder shall be dissolved in plenty of water before it is spread in the trench.

12.2.2.4.4. Inspection

Video cameras and other equipment inserted into pipelines shall be disinfected as follows. Cameras shall be dunked in a bucket of HTH solution, 1% w/w for 20 seconds, and the cable shall be fed through a rag soaked in the same HTH solution.

12.2.2.4.5. Disinfection of personnel working inside pipes

People working inside pipes shall wear white rubber boots and clean disposable overalls. Footwear shall be disinfected by people placing their feet in a bucket of HTH solution, 1% w/w for 20 seconds. People exiting pipes shall remove overalls and put on fresh overalls if they re-enter the pipes.

12.2.2.4.6. Disinfection of equipment

Small equipment shall be disinfected by immersion in a large bucket of HTH solution, 1% w/w. Welding leads shall be fed through a rag soaked in the same HTH solution.

Larger equipment, such as submersible pumps, hoses, gas bottles, etc., shall be disinfected by spraying with a 1% sodium hypochlorite solution.

12.2.2.4.7. Disinfection of pipe at repair

12.2.2.4.7.1. Splicing in a new section

12.2.2.4.7.1.1. Pipe 100–700 mm diameter

These pipes shall be disinfected by placing 200 g of HTH in the pipe section to be spliced in.

12.2.2.4.7.1.2. Pipe 750 mm diameter and over

The whole inside area of the pipe to be spliced in shall be sprayed with a 1% NaOCl solution.[23]

12.2.2.4.7.2. Repairs inside an accessible pipe. Only the area repaired shall be sprayed with a 1% NaOCl solution.

12.3. ISO 9001 Policy (suggested)

- The pipeline system shall be designed, constructed and operated so that all treated water meets the required water quality.
- The pipeline system shall be designed, constructed and operated to prevent contamination from outside sources.

The design of the pipeline system will include structural considerations, materials (pipe and linings) and the practical aspects referred to in the procedures above. Additional areas of concern for design are spare branches, cross-connections, air valve and scour valve arrangements, and the elimination of silt/debris traps.

12.4. Disinfection of new pipelines

For a full treatise on this subject, reference should be made to AWWA C601.81: *Disinfecting water mains*. Disinfection of new pipelines is part of the construction work before a pipeline is commissioned. This work normally follows the acceptance of the pressure test. The steps are as follows.

1. Remove contamination by flushing or manually cleaning the pipeline.
2. Distribute the chlorine (e.g. HTH granules) throughout the pipeline.
3. Fill with water and leave for 24 hours.
4. Dechlorinate and drain.
5. Recharge the pipeline and test for coeliform bacteria at several sample sites.
6. Open valves to connect new pipeline to the existing pipeline after negative bug test results.

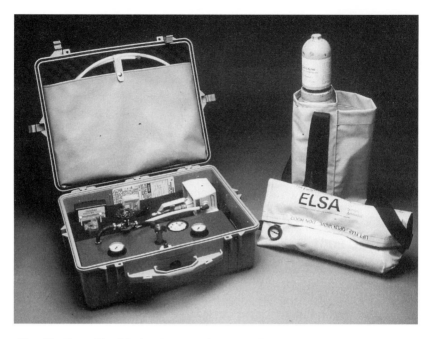

Fig. 48. Portable chlorination unit (courtesy Design Water, New Zealand)

The chlorine level required for disinfection is 25 ppm in AWWA C601.81, but up to 50 ppm may be used. The method of chlorination can be liquid (sodium hypochlorite), solid (HTH granules or tablets) or gas. Direct dosing of a large long pipeline with a dose pump will be too slow. A portable gas chlorinator with testing of the chlorine level using a colour comparator gives a homogeneous chlorine solution (refer Fig. 48). Portable gas chlorinators can also be used in an emergency situation to disinfect an alternative water supply, say from a local stream. The simplest method is to use HTH granules or tablets, strategically dispersed throughout the pipeline system. The HTH can be added at air valve branches or other access points. Water from the local town supply can be used to fill the pipeline.

A challenge is posed by the draining of the pipeline after a successful disinfection. The superchlorinated water can be drained into the local sewer system so that the chlorine is used up. It is possible to drain into a watercourse but this necessitates dechlorination and a dose pump using the strongest dechlorination chemical solution possible. The drain rate could be controlled to allow dechlorination to take place within the parameters of the normal dechlorination equipment (refer 5.5.4.1). For example, a dose pump capable of neutralising 500 l/s of treated water at 0·7 ppm, chlorine residual will neutralise 7 l/s of superchlorinated water at 50 ppm chlorine residual.

13. Health and safety issues

The last three decades have seen great changes in the approach to Health and Safety at work. This chapter is included because Health and Safety is an integral part of the work on bulk water pipelines. The Pipeline Authority in its role as principal is responsible and legally liable for harm to employees or subcontractors caused by bad working practices or negligence.

The emphasis today is on *hazard management*. The hazards associated with and arising from the activities of pipeline construction, maintenance and operation have to be identified and then addressed with the full involvement and co-operation of the workforce. Health and Safety should be a design perspective, not simply a response to the existing work situation. For example, a shallow valve chamber with a 600 mm diameter single access/egress point will be classed as a confined-space and will require a dedicated safety control plan. If the chamber roof were designed to have removable lid sections, the access and ventilation would be improved so that the installation would not be a confined-space. The safe system of work when ongoing maintenance work was being carried out would then be much simpler.

The typical work on bulk water pipelines is similar and, in many cases, identical to the activities of a construction company, so it can be seen that the Health and Safety perspective demands considerable time and energy from the Pipeline Engineer and field supervisors. It should be noted that hazards affect not only the workers involved in specific activities but also the co-workers, passers-by, visitors and members of the public.

The hazards identified from a 'brainstorming' meeting of workers, supervisors and engineers are catalogued, and each one of them is then analysed. The parameters gathered for each hazard are the risk (based on frequency of occurrence) and consequences (e.g. injury, serious injury or death). The long-term nature of some hazards must also be recognised, e.g. noise, fumes, etc. The hazards catalogued are then judged to determine whether or not action is required, which is based on *reasonable steps* being taken by the principal. Hazards requiring action, i.e. management, can be addressed in three ways.

1. *Eliminate the hazard*
   A new procedure/working practice is adopted: for example, the repairing of lead jointed cast iron pipes. The old method of pouring lead into the socket is abandoned and a cut and splice method (refer 1.7) or other mechanical methods (refer 3.8) are used. The new procedure becomes the ruling working practice and is referenced and detailed in a hazard control plan.

2. *Isolate the hazard*
   The present procedure/working practice is changed so that the hazard is always physically isolated from the worker: for example, the blanking of valves before internal inspection of large diameter pipelines is carried out (refer 5.1). This prevents water from entering the confined space.

3. *Minimise the hazard*
   The worker is protected from the hazard, but the hazard is still present: for example, a welder must wear protective clothing — welding visor, gloves, jacket, trousers, ear plugs, safety footwear — and fume extraction equipment must be used in a workshop setting.

Hazards that are considered not to be significant do not require any action. Significant hazards that can lead to injuries will require action and control. The hazard control plan can be audited to prove that the principal has acted correctly at all times. The control plan will include regular testing of equipment and certification. Specific safety training will also be nominated and records kept of qualified personnel. The following control plan for hot work is given as an example.

HOT WORK HAZARD CONTROL PLAN
1. *Use/Occurrence/Activity*
   Oxy-acetylene cutting/welding sets
   LPG heating equipment
   Electric welding

2. *Hazards*   *Possible consequences*

Explosion	Burns, injuries, possibly death
Fire	Burns, asphyxiation
Fumes	Asphyxiation
Unsecured load	Injuries
Sparks	Eye injuries, burns

3. *Control*

Only suitably trained and equipped personnel shall use oxy-acetylene cutting/welding sets and electric welding sets.

The following detailed guidelines,[53] courtesy of Wellington Regional Council, are appended to the Control Plan as extra material. Other safety material such as manufacturers' safety guides and government publications can also be appended.

Hazards

Hot work includes gas cutting, torch welding, arc welding, brazing and soldering.

The main hazards from these operations are fire, burns, (flame, surface contact or radiation), toxic fumes and eye damage. Hot work on enclosed vessels such as drums and tanks may give rise to ignition of their internal atmosphere. Even substances not normally considered flammable at room temperature (e.g. grease) may explode under such circumstances.

Hazard management

Options to hot work should always be considered, such as manual cutting or cold soldering (adhesives and fillers).

Hot work in a confined-space is a specialist operation and must not be undertaken.

Only suitably trained and equipped personnel may use gas cutting and welding equipment.

Flashback arrestment devices must be fitted to gas bottles used in cutting and welding operations.

All equipment must be checked thoroughly before use. In particular, gas hoses must be checked for leaks and the insulation on arc welding sets must be checked. Note: check for gas leaks with a detergent/water mix (not soap).

No oil or grease may be used to lubricate the threads on oxygen cylinders, as high pressure oxygen may react explosively with it.

Acetylene bottles must never be stood more than 45 degrees from their upright position. This is because the liquid acetone within may foul the valve assembly and cause a dangerous situation.

When gas cylinders are being transported other than on an appropriate trolley, valve caps must be fitted and a check made that the cylinders are securely mounted.

Compressed gas cylinders must not be dragged, bumped, rolled, heated or otherwise damaged.

Acetylene reacts explosively with copper and is unstable at pressures greater than 100 kPa.

Empty gas cylinders must be treated with the same caution as full ones as they may still contain an explosive residual atmosphere.

A fire extinguisher must always be available for hot work. Sparks from gas cutting operations may be an ignition hazard within a 10 m radius of the work. Conduction of heat may also cause ignition of adjacent combustibles. All combustible materials within this radius should be removed, shielded or otherwise protected.

Screen torch operations to reduce the risk of fire and to avoid eye irritation to those nearby. When arc welding is being undertaken, protection is required for all those within 15 m of the work.

Overalls, gloves, apron, eye protection, filter respirators and long boots are required to be worn for welding operations. Overalls should fit tightly at the neck and wrist. Nylon overalls or jackets and clothing with exposed pockets may not be worn. Goggles must be worn for gas cutting and welding operations and a face shield or visor for arc welding. Head protection is optional. Gloves or gauntlets are necessary during arc welding to protect against shock, burns and radiation exposure.

A torch, whether lit or otherwise, should never be pointed towards clothing, as gas trapped in the fabric of clothing can cause severe burns if ignited.

Hot welding rod stubs, slag and off cuts should be disposed of in a safe manner, such as into sand.

Welding and cutting operations generate toxic metal fumes including ozone, nitrogen dioxide and carbon monoxide. Such work must be carried out in a well-ventilated area, preferably by means of a local exhaust ventilation unit. A welding respirator must always be worn by the operator.

Brazing, in particular, produces corrosive and toxic fumes, while the vaporisation of grease, paint and other coatings on the metal will add further components to the fumes generated. Brazing with cadmium based rods presents a high risk to the user and local exhaust ventilation to remove fumes at source will always be required.

Electric arc welding units must have an effective earthing arrangement using an earthing clamp or bolted terminal, which should be checked regularly. Arc welding must not be attempted if hands are wet or floors are damp.

An isolating transformer should be used for arc welding.

Before hot work is started, the operator must be satisfied that the equipment is in good working order and that the surrounding environment and other people are protected from hazards which may arise from the proposed work. This includes screening of ultraviolet radiation emissions from arc welding.

Table 5. List of chemicals that are hazardous when heated

Acetone	Acrylonitrile	Amyl acetate
Amyl alcohol	Ammonia and compounds	Calcium hypochlorite
Carbon disulphide	Chloroform	Sodium perchlorate
Sodium chlorate	Potassium perchlorate	Potassium chlorate
Cyanides	Dichloromethane	Formaldehyde
Hydrochloric acid	Isopropyl alcohol	Gylcol ethers
Glycol ether acetate	Methyl alcohol	Methyl ethyl ketone
Mineral turpentine	Nitric acid	Phenol
Styrene	Sodium hypochlorite	Hydrocarbon solvents
Trichloroethane	Toluene	Sulphuric acid 1.1.1
White spirit	Ammonium perchlorate	Xylene
Turpentine	Ammonium chlorate	Kerosene

A barrel or container holding hazardous or flammable material residues should never be welded or cut, as a flammable or toxic atmosphere may be generated during the heating process. Flammable vapours may arise from:

- volatile liquids
- non-volatile oil, tar, grease, soap
- acids which react with metal to form hydrogen
- combustible solid powders (e.g. fibre glass, sulphur, milk powder).

In Table 5, some of the more commonly encountered substances stored in drums or tanks, which may be hazardous when heated, are listed.

Explosive dusts
It is wise to treat all dusts as potentially explosive.

14. Engineering drawings

Engineering drawings represent on paper the physical bulk water system, and therefore constitute valuable intellectual property. A drawing system is usually a curse or a blessing to use. A logical, easy to use system, updated whenever a change takes place, should be the goal. Up-to-date working sets of drawings in the depots (for daily use and reference), in the field (significant drawings for daily use) and in the head office (reference set) is the objective.

The structure of a drawing system is a contentious issue but the following is nevertheless offered.

14.1. Drawing numbering system

1. Block Identification
This label is a unique alphanumeric for each block (subsystem). New projects within a block generate drawings under *that block*.

2. Discipline Identification
Separate alphabetical centres for each discipline, e.g.

C for Civil
E for Electrical
M for Mechanical
P for Piping.

3. Numbers for each Drawing (note: number alone is not unique),
e.g. C1001, C1002..., P1001, P1002..., etc.

14.2. Drawing types

Engineering drawings range from the schematic to the detailed, from the general overview (e.g. general arrangement) to the specific (e.g. a pressure gauge fitting). There is an acknowledged hierarchy of drawings as follows:

- flow scheme
- general arrangement
- detail drawings
- 'finder' drawings.

14.2.1. Flow schemes

The flow scheme is a logic diagram whose purpose is to mimic the actual system. Therefore, it is the most important drawing from an operational point of view.

The flow scheme is the 'bible' of the process under consideration. A small to medium size water system might fit on one sheet, with several sub-drawings showing the smaller chunks of process. A large water system will need a number of drawings which might need to be mounted on a big wall to show the complete picture.

A true flow scheme will show pipe sizes, flows and instrumentation logic. For waterworks purposes, the flow scheme should show *all* the main valves and any parallel pipelines. It can also depict reservoir capacities and top levels. A quick look at such a drawing will reveal if the flow direction can be reversed on a main in an emergency. One higher elevation reservoir could supply a lower elevation reservoir.

14.2.2. General arrangement

This type of drawing is self-explanatory. Typical examples are pump station pumps and pipework, and valve chamber pipework arrangements. They show the gist without all the detail.

14.2.3. Detail drawings

The detail drawing shows all the dimensions of the item under consideration so that a person can construct the pipework shown.

14.2.4. Finder drawings

The finder drawing is a schematic drawing which shows valve and other equipment locations. The value of the finder is that it shows a great deal of pipeline on one drawing. This avoids having to carry a large number of as-built drawings in the field. Finders are derived from as-builts so, if in any doubt, believe the as-built not the finder.

14.3. Review of drawings

A thorough systematic review of the drawings is advisable once every 10 years or so. The information on the drawings should be cross checked against the reality of the pipelines in the field, wherever possible. Further reviews may not be necessary *provided* that field changes are accurately surveyed and recorded.

14.4. Geographic Information System

The Geographic Information System (GIS) is more than a conventional drawing system. It does not use local co-ordinates to define the position of pipelines or other features. The co-ordinates are global, and field positions are determined by way of satellites. The main use of the GIS is as an Asset Management tool.

The GIS will typically show roads, rivers, houses and some utilities' lifelines. It also has the capability to store information on long sections and specific details. A CAD drawing system can feed into the GIS. The accuracy of the GIS is now quite good, ±500 mm, so that one could confidently locate the position of a bulk water pipe in the field.

Conventional drawing information can be retrojected into the GIS system and verified by checking a number of actual field positions, e.g. valves, bends. In time, GIS systems can be expected totally to absorb conventional drawing systems, including CAD, as GIS alone gives an integrated 'picture' of a pipeline's habitat.

15. Pipelines in land

The land in which water pipelines are situated is of vital concern to the water authority. The critical perspective is one of having legal rights to maintain the pipeline concerned. This is more straightforward for owned land. All other land through which the pipeline passes requires an easement or, where this may not be granted, a lease with a 'good reason to move pipe' clause built in. Various Acts of Government may allow authorities to repair water pipelines in case of emergency, regardless of the lack of easements. The legal situation will vary from country to country so my approach is a list of critical questions.

- What right does the water authority have to construct pipelines on land occupied by others?
- How wide should the pipeline access/maintenance strip be?
- What is the tenure of the land over or through which the pipeline passes?
- Are there any particular difficulties with particular land tenures?
- What is the best method to ensure access for maintenance and replacement is in perpetuity?
- What alternative routes are there?
- What are the costs in obtaining access rights?
- Are there any on-going costs associated with the access rights?

A definition of an easement follows: 'An easement is a right attached to land which permits the owner of the dominant tenement (the pipeline owner) to require the owner of the servient tenement (the land owner) to accept a restriction on the land. Easements can come in many forms — surface, sub-surface, air rights . . .'.[54]

Glossary

AC	Asbestos cement
Access point	450 mm or 500 mm diameter short flanged branch allowing access to the interior of a pipe
Air valve	Valve allowing air to escape the pipeline and/or to enter the pipeline to prevent a vacuum
ANSI	American National Standards Institute
Aqueduct	*see* Bulk water pipeline
Auxiliary supply	*see* Private supply
AWWA	American Waterworks Association
Backflow preventer	Minimum specification: a series of two non-return valves with test valves to facilitate testing of the backflow prevention function. The backflow preventer stops backflow of water into the main pipe when pressure is reduced, e.g. when the main is emptied for maintenance.
Bar	$10^5 \, \text{N/m}^2$
Blowoff valve	*see* Scour valve
BS	British Standard
Bulk water pipeline	Treated water pipeline connecting treatment plants to city/town reservoirs; also, raw water pipelines from intakes to treatment plants
CAD	Computer aided drafting
Check valve	*see* Non-return valve
Coating	Corrosion-resistant layer of material on the outside of a pipe
Coupling	Means of joining two pipes together using an extra element
CP	Cathodic protection
CTE	Coal Tar Enamel
Cut-ins	Connection of an existing pipe to a new pipe
DAV	Double air valve
Deviation	New pipe laid between two points on the old line to avoid an obstacle
Drain valve	*see* Scour valve

Easement	Legal right to access a pipeline sited on another's land for construction/maintenance and operation purposes
EFW	Electric fusion welding
EPDM	Ethylene propylene diene monomer
GIS	Geographic information system
GRP	Glassfibre reinforced polyester
Harness	Means of providing axial restraint across a joint; requires lugs welded to both sides of the pipe to locate tie-bolts
HTH	High test hypochlorite (Calcium hypochlorite)
Joint	Generic term meaning the joining of two pipes together
kPa	Kilopascal $10\,N/m^2$
Line valve	*see* Main valve
Lining	Corrosion-resistant layer of material on inside of pipe. Lining must not affect the quality of the drinking water
m	Metre — surrogate unit of pressure, *metres head*
Main	*see* Bulk water pipeline
Main valve	Valve isolating a section of main pipeline; can be actuated by electric motor or other means of actuation
MDPE	Medium density polyethylene
Ml	10^6 litres
MMA	Manual metal arc welding
MPa	Megapascal $10^6\,N/m^2$
NRV	Non-return valve
NZS	New Zealand Standard
PE	Polyethylene
Pigging	Method of cleaning pipes by sending a swab (pig) down the line, either by existing water pressure or by compressed air; it is important to pig newly constructed pipelines before commissioning
Private supply	Small connection on a bulk main, supplying an individual property directly
RCW	Regional Council Wellington, New Zealand
Reflux valve	*see* Non-return valve
RRJ	Rubber ring joint (spigot and socket); can be found on most pipe materials
SAV	Single air valve
Scour valve	Valve fitted on bulk mains to allow draining of water, usually installed at low points
Special	Custom made pipe fabrication (spool)
Spectacle blind	Open ring and closed circular plate joined together, and inserted between two flanges to provide a safe shut off; normally fitted in the open position
Tie-bolts	*see* Harness
Tie-in	*see* Cut-in
TIG	Tungsten inert gas welding
TMP	Traffic Management Plan

Tomming	Bracing of bends, tees and dead ends against excavation walls to resist thrusts
Trunk main	*see* Bulk water pipelines
uPVC	Unplasticised polyvinylchloride
USEPA	United States Environmental Protection Agency
Valve chamber	Buried structure which allows access to valves for maintenance purposes
Weld band	Split socket which is welded to the outside of a pipe with two fillet welds after assembly through the tightening of a tension bolt

References

1. de ROSA P.J. *Pipeline materials selection manual*. Water Research Council, 1988.
2. DRESSENDORFER P.V. and HALFF A.H. Large water mains: experience and practice of three large users. *American Waterworks Association (AWWA) J.*, 1972, July, 435–440.
3. HOLSEN T.M. *et al*. The contamination by permeation of plastic pipe. *AWWA J.*, 1991, Aug.
4. Getting a grip on pipes. *Water & Environment*, 1996, July, 16.
5. SCHIFF M.J. and McCOLLOM B. Impressed current cathodic protection of polyethylene-encased ductile iron pipe. *MP*, 1993, Aug., 23–27.
6. SWICHTENBERG W. Pipe expansion project quickly increases production capacity. *Water Engineering and Management*, 1995, Oct., 28, 29.
7. Light mastic smoothes joints. *World Water & Environmental Engineering*, 1993, March/April, 54.
8. de ROSA P.J. *Op. cit.* 130.
9. PVC pipe design and installation. *AWWA J.*, M23, 1980, 77.
10. PHILLIPS R.V. *et al*. Pipeline problems: brittle failure, joint stresses and welding. *AWWA J.*, 1972, 421–429.
11. AMERICAN WATERWORKS ASSOCIATION. *Specification for lock bar pipe*. AWWA, 1942. Tentative Standard 7A.2.
12. AMERICAN WATERWORKS ASSOCIATION. *Steel pipe design manual* M11. AWWA, 1989, 1.
13. TUTHILL A.H. Stainless steel piping. *AWWA J.*, 1994, July, 67–73.
14. YOUNG M.B. and MEANS E.G. III. Earthquake lessons pay off in Southern California. *AWWA J.*, 1995, May, 59–64.
15. Ductile iron pipelines. Tubemakers Australia, 1986, 1–3.
16. PLISHKA M.J. and SHENKIRYK M. Chemical cleaning process for water systems. *Water Engineering and Management*, 1996, Mar.
17. KLEIN R.L. and RANCOMBE A.J. Performance of water pipeline materials. *Chemistry and Industry*, 1985, 355.
18. Epoxy repair compound for cast iron pipes. *World Water & Environmental Engineering*, 1996, Dec.
19. Polyethylene pipe, air lift at Norwegian Fjord. *World Water & Environmental Engineering*, 1994, Mar.

20. CHAU K.W. and NG V. A review of the design practices of thrust blocks for water pipelines in Hong Kong. *Aqua*, **45**, No. 2, 92–95.

21. BRITISH STANDARDS INSTITUTION. *Specification for Class II arc welding of carbon steel pipework for carrying fluids*. BSI, London, 1991, BS 2971.

22. BALLANTYNE D. Minimising earthquake damage. *World Water & Environmental Engineering*, 1995, Sept.

23. AMERICAN WATERWORKS ASSOCIATION. *Disinfection of water mains*. AWWA, 1981, Standard C601.81, 9.

24. de ROSA P.J. *Op. cit.* 127, 128.

25. *Hobas Design Textbook*. Hardie Iplex, 1994, 2–7.

26. German technology. *WWI*, 1995, Oct., 20.

27. STEPHENSON D. *Pipeline design for water engineers*. Elsevier, 1976, 182.

28. Distribution system maintenance techniques. *AWWA J.*, 1987, 19.

29. Pipe freezing technology. *World Water & Environmental Engineering*, 1994, June.

30. AMERICAN WATERWORKS ASSOCIATION. *Distribution valves: selection, installation, field testing and maintenance manual* M44, AWWA, 1996.

31. CHADWICK A.J. and MORFETT J.C. *Hydraulics in civil engineering*, Allen & Unwin, 1986, 209.

32. Control system closes in on leaks. *Water & Environment*, 1996, Mar., 38.

33. MACLELLAN D.A.S. Ventilation of pipeline systems and the selection and application of air valves. *Pipes and Pipelines International*, 1965, May/June, **10**, Nos 5&6.

34. Air in pipelines, sources, system impact and removal. Valmatic, 1993.

35. Air valve technology reviewed. *Vent-O-Mat*, 1996, Jan.

36. AIKMAN D.I. Asset appraisal of trunk mains. *IWEM*, 1993, Feb., 42.

37. DVIR Y. *Flow control devices*. Control Appliances Books, Israel, 1995, 369–384.

38. SINGLETON J. AMERICAN WATERWORKS ASSOCIATION. Dechlorination unit improves safety, efficiency. *Opflow*, 1993, May, AWWA.

39. CLA-VAL COMPANY. *Cla-Val automatic valves*. Cla-Val Company, 1971, 29.

40. WELLINGTON REGIONAL COUNCIL. *Specification for the provision of cathodic protection for water pipelines*. Parts A and B. Wellington Regional Council, 1997.

41. D'ATH R. *Cathodic protection of buried pipelines*, New Zealand Water and Wastes Association, 1996, 165.

42. CRITCHLEY R.F. and AIKMAN D.I. Aqueduct management planning: Thirlmere, Haweswater and Vyrnwry aqueducts, *J. IWEM*, 1994, Oct.

43. RUSSELL D., FERGUSSON P., HEATHCOTE M. and MOORE G. Condition assessment of water mains using remote field technology.

44. The tiniest robots in the world. *The Weekly Telegraph*, 1997, No. 299, April.

45. GLASER K.C. AMERICAN WATERWORKS ASSOCIATION. OSHA releases new confined-space safety regulations. *Opflow*, 1993, June, 5.

46. BLAINE B.A. Proc. IWEM Symp. on 'Leakage Control in the Water Industry'. *J. IWEM*, 1993, **7**, Oct.

47. Control system closes in on leaks. *Water & Environment*, 1996, Mar., 38.

48. GRIMAUD A. and PASCAL O. Monitoring system detects small leaks. *Water/Engineering & Management*, 1991, Jan., 14, 15.

49. WATER RESEARCH COUNCIL. *Manual for MDPE pipe systems for water supply*, WRC, 1986, Part 8.

50. *Construction Act New Zealand.* Government Print, 1959.
51. de ROSA P.J. *Op. cit.* 30, 31.
52. AMERICAN WATERWORKS ASSOCIATION. *Distribution system maintenance techniques.* AWWA, 1987, 82–85.
53. WELLINGTON REGIONAL COUNCIL. Health and safety guidelines: workshops and storage. Wellington Regional Council, 1995.
54. LARMER J. Gas and oil pipeline easements on farmland (valuation considerations). *New Zealand Valuer's J.*, 1995, Dec.
55. ROBERTS D. and BRADFORD P. A practical approach to risk assessment in Essex and Suffolk Water. *Aqua*, **45**, 1996, 213–218.

Bibliography

AMERICAN WATERWORKS ASSOCIATION. *Mechanical sleeve couplings.* AWWA, 1991, AWWA Standard C219–91.

BACKMANN W. and SCHWENK W. *Handbook of cathodic protection.* Portcullis Press, 1975.

BRITISH STANDARDS INSTITUTION. *Specification for flanges and bolting for pipes, valves and fittings.* BSI, London, 1962, BS 10.

BRITISH STANDARDS INSTITUTION. *Circular flanges for pipes, valves and fittings (PN designated). Part 3: Section 3.1. Specification for steel flanges.* BSI, London, 1989, BS 4504.

BRITISH STANDARDS INSTITUTION. *Cathodic protection. Code of practice for land and marine applications.* BSI, London, 1991, BS 7361: Part 1.

FEBCO. *Cross connection control handbook.* 1977.

INTERNATIONAL ORGANISATION FOR STANDARDISATION. *Quality Systems International Standard.* ISO, 1987, ISO 9000.

INTERNATIONAL ORGANISATION FOR STANDARDISATION. *Specification for design/ development, production, installation and servicing.* ISO, 1987, ISO 9001.

INTERNATIONAL ORGANISATION FOR STANDARDISATION. *Specification for production and installation.* ISO, 1987, ISO 9002.

JANSON L.A. *Plastic pipes for water supply and sewage disposal.* Borealis, 1995.

STANDARDS NEW ZEALAND. *Specification for welded steel pipes and fittings for water, sewage and medium pressure gas.* NZS, Wellington, 1988, NZS 4442.

PARKER M.E. and PEATTIE E.G. *Pipeline corrosion and cathodic protection.* Gulf Publishing, 1984.

Appendix
Earthquake response plan

Author's Note

This material, while specifically addressing the response of water authorities to earthquakes, does have some relevance to non-seismic areas, e.g. in the analysis of vulnerable elements of the bulk water system, such as pipeline crossings.[55] It is based on the author's work in 1994 as part of the Wellington Lifelines Sub-Group, looking at earthquake response planning for water supplies. The Lifeline Group consists of personnel from all the lifeline utilities in Wellington, New Zealand, who meet regularly to progress the integration of earthquake response planning in the Wellington area.

Three Sub-Group members developed a generic response plan containing all the elements under three phases: initial response; recovery of partial service; and recovery of full service. For a large bulk water supply with many cross-connections, recovery of partial service is a valid idea. For a smaller system, however, recovery of partial service may be a less valid concept, as the system either works or not.

Some benefits of an earthquake response plan are as follows:

- a better understanding of the bulk water system and the vulnerable parts
- an estimate of the mitigation measures must be made and planned/funded in the business plan; likewise the preparedness measures
- a reduction in the risk to the system and increased confidence in the performance of the system. Confidence in the capacity to respond and repair/recover the system.

A generic bulk water earthquake response plan follows, with some specifics for illustrative purposes. Individual authorities can add their own specific details and tailor the plan to suit their own situation.

Contents

Introduction

What To Do

All staff have separate detailed instructions of what to do and where to report. Make sure you know what to do. If in doubt, ask your supervisor. Always carry with you your Civil Defence identification card.

The time, day of week and particular circumstances will determine how each individual is placed at the instant of an earthquake. Each person is to be mentally prepared for the different possibilities.

As a guide to ground motion caused by a damaging earthquake, it has been determined that ground movement above Level VII on the Modified Mercalli Scale is likely to cause widespread damage. The indications of this degree of ground motion are as follows:

- There will be general alarm which may approach panic.
- It will be difficult to stand.
- Heavy furniture will be overturned.
- People who are driving cars will notice that the steering is affected.

Initial actions

Most staff have families and will be concerned to know what has happened to them. No hard and fast rule can be laid down. However, it should be remembered that each person's particular skills and experience are invaluable in restoring order after an earthquake. You have a responsibility to the community as well as to your family.

Talk about these matters with your family and let them read these instructions so that they will not be unduly alarmed if you do not return home for some time after an earthquake.

Action List

The Action List should comprise instructions such as:

- Manager and engineers to report to the emergency operation centre
- Pipeline and treatment plant staff to report to (normal depot)
- Treatment plant staff to inspect all equipment to a predetermined list

Emergency Response Plan

1.1. Objective and scope of plan

The objective of this plan is to facilitate a pre-planned, yet flexible, *response* to a major earthquake event.

The scope of the plan is the total response to the emergency event, that is, initial response, recovery of partial service and the restoration of full bulk water service. It applies to the whole of the bulk water system.

The scope of this plan also includes pre-event mitigation and preparation actions.

Annual audit

An annual audit is to be undertaken by an external consultant to ensure that the Emergency Response Plan is up to date and to check on any actions which have been proposed during the particular year of the audit.

Annual exercise

An annual exercise is to be conducted involving all staff to ensure that they are well-prepared for an emergency.

1.2. Summary of organisation and management structure

The core management group is comprised of the following personnel:

- *Define personnel and their roles*

1.2.1. Overview of actions to be taken in the event of an emergency

The actions to be taken in the event of an emergency are:

1. to inspect the system according to a predetermined hierarchy.
2. to transmit the inspection information to the emergency operation centre.
3. to inform the Regional Civil Defence Headquarters of bulk water supply status.
4. to keep the system operational where possible.
5. any specific operational actions (*define*).

1.2.2. Emergency operation centre

1.2.2.1. Activation

The most senior member present at the emergency operation centre will act as the Duty Officer until a more senior staff member arrives and becomes Duty Officer. This substitution will continue until the designated core management team arrives.

1.2.2.2. Location

The emergency operation centre is located at (*define street names, etc.*). The most senior staff member will determine the operational status of the emergency operation centre.

This acting Duty Officer will declare the defined emergency operation centre operational, or if unusable, declare (*define back-up emergency operation centre location*) as the emergency operation centre.

1.2.2.3. Functions

The functions are:

1. to act as an information gathering centre
2. to assess the nature and extent of the emergency
3. to direct and co-ordinate personnel, materials and services
4. to co-ordinate assistance provided by other local authorities within the Region and agencies outside the Region
5. to inform and keep informed the Regional Civil Defence Headquarters of the current situation and the water authority's requirements for assistance.

1.2.2.4. Emergency provisions

* *Define food, water and sleeping provisions and the duration of the provisions.*

1.2.3. Management structure

1.2.3.1. Structure

The core management group will work as two teams, each working a 12 hour shift. As personnel become available, each team will consist of a Duty Officer and a deputy, as a minimum.

1.2.3.2. Function

The function is:

the operation and repair of the bulk water system.

Each team will communicate regularly with the Regional Civil Defence Headquarters on the status of the bulk water system and any requirements for assistance.

1.2.3.3. Authority

* *Define the financial authority each team/person has.*

1.2.3.4. Organisation chart

Figure 49 depicts an organisation chart for emergency reponse.

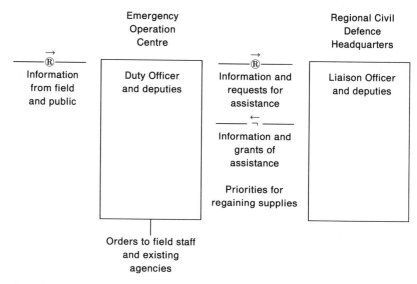

Fig. 49. Organisation chart for emergency response (courtesy Wellington Regional Council, New Zealand)

1.3. Interfaces with other authorities
1.3.1. Civil Defence
1.3.1.1. Responsibility boundaries
- *Define your authority's responsibilities. Are you responsible for providing alternative supplies?*

1.3.1.2. Liaison
The Liaison Officers are based at the Regional Civil Defence Headquarters and provide the link between the core management based at the emergency operation centre and Regional Civil Defence.

1.3.2. Clients (e.g. for water wholesalers)
The Duty Officer will establish contact with the person(s) responsible for water supply as soon as possible after the event.

- *Define your clients and contact details.*

1.3.3. Other utilities
The Duty Officer will establish contact with the person(s) responsible for the following services when necessary:

- Power
- Gas
- Rail
- Telecommunications
- Highway Authorities.
- *Define contacts and details*

1.4. Immediate response phase
1.4.1. Activation
This plan is activated by the emergency itself. Declaration of a state of emergency will be given by a series of short blasts on hooters, loud-hailers, police car sirens or any other alarm device. This will be supplemented, if practicable, by radio broadcasts giving civil defence information and instructions.

1.4.2. Establishment of emergency operation centre
The most senior staff member at the emergency operation centre will act as the Duty Officer until a more senior staff member arrives. This substitution will continue until the designated core management team arrives.

1.4.3. Communication arrangements
1.4.3.1. Internal
Vehicles will each have radiotelephone sets with telecommunication interfaces incorporated. Communication by way of radiotelephone can be to other vehicles and bulk water bases, e.g. emergency operation centre.

1.4.3.2. External
Communication is by telephone only at present.

- *Define external communications.*

1.4.4. Inspection hierarchy
- *Define the 10 most vulnerable parts of the water system.*

1.4.5. Safety procedures / precautions
The safe approach to the inspection of, and later working on, pipelines and equipment damaged by an earthquake is firstly to do a *hazard analysis*. The following hazards are an indication only, not an exhaustive list:

- slips and falling rocks/trees
- buildings in danger of collapse
- explosion/fire from escaped gas
- electrical cables.

All the above hazards can be managed by avoiding the edge of slips, by not entering buildings that show damage and by avoiding areas that smell of gas or where the sound of escaping gas is audible.

1.4.6. Receive and record information
- *Define the information system including information flow paths.*

1.4.7. Analyse information
The Deputy Officers will produce the following lists of:

1. all reservoirs still receiving a bulk supply
2. all main valves which will isolate damaged sections
3. alternative supply options from within the system
4. materials and equipment required to repair the system
5. staff required to repair the system.

1.4.8. Interfaces with clients

There will be a meeting with clients to identify their priorities. This meeting will take place within 24 hours after the emergency event.

1.4.9. Set priorities for response and control

The Duty Officer will set priorities for operating the bulk water system and repairing the bulk water system.

The Deputy Officers will produce plans detailing these priorities and showing the staffing levels, materials and plant required for each repair.

1.4.10. Review ability to respond

The Duty Officer and deputies will review the available resources against the estimated resources detailed on the plans.

1.4.11. Implementation of mutual aid agreements

The Duty Officer will implement any mutual aid agreements needed to help meet the initial priorities.

1.4.12. Need for alternative supply options

The analysis of the situation will show which reservoirs are not supplied with water. The Duty Officer may authorise use of the following water supplies where necessary.

* *Define hydrants on the bulk water pipelines.*

1.4.13. Media management / public information

1.4.13.1. Emergency Public Information Centre (EPIC)
* *Define your specific arrangements.*

1.4.13.2. Direct media communications

In certain instances the media may seek further details on information issued by the EPIC and approach the emergency operation centre directly. In this case, one nominated person from the management team on duty will speak to the media on behalf of the bulk water authority.

Information given to the media should give the following details of the pipeline repairs:

* Where and when the repairs are likely to be finished (timing given should be conservative)
* What areas of the Region are affected by the repairs
* How many people are working on repairs.

1.4.13.3. Communication by members of the public
* *Define how communication by the public will be managed.*

1.5. Recovery of Partial Service
1.5.1. General

The Recovery of Partial Service Phase is to effect a minimum service level around the Region. This takes account of available and recoverable supply point and pipe networks. It typically addresses up to *two weeks* after a major event.

1.5.2. Organisational set-up

Actions will be based on decisions made by the Duty Officer after consultation with client water supply managers.

1.5.3. Implementation of mutual aid agreements

The Duty Officer will implement any further mutual aid from existing agreements.

1.5.4. Provision of pipeline materials

• *Define emergency backup materials available and locations.*

1.5.5. Record repairs

All repairs shall be recorded on the as-built pipeline drawings. A sketch of the repair shall be forwarded to Head Office showing the following:

• date of repair and name(s) of welder(s)
• dimensions of any specials made
• details of connections, e.g. welding bands or couplings
• anchoring/support details
• repairs to the internal lining
• repairs to the external coating.

1.5.6. Internal inspection

Pipelines large enough to be inspected internally shall be inspected fully for damage and shall be repaired accordingly. Pipelines are confined spaces. The Health and Safety Guidelines for entering/working in a confined space shall be followed.

On completion of repairs, the start-up procedures/valving of Appendix 1.5.7 shall apply.

1.5.7. Start-up procedures / valving

Repairs will have begun on pipeline sections downstream of isolated valves. On completion of a repair, the bypass valve shall be opened slowly to charge the repaired section up to the next line valve. When the noise on the bypass valve ceases, the main valve shall be opened.

If the bypass valve continues to roar after an hour, then the section of pipeline downstream shall be inspected for further damage.

If the pipeline section is charged and holding pressure, then the procedure repeats for the next section once any repairs are complete.

1.5.8. Safety procedures

In addition to the procedures of Appendix 1.4.5, the usual safety procedures apply.

Before work is carried out on pipelines, reference should be made to Appendix 1.9.7 (Excavation procedures).

1.6. Restoration of full service
1.6.1. General

The restoration of full service phase is to provide an ordered programme to return to full service as soon as possible.

It typically addresses two weeks to six months or more after the event.

1.6.2. Organisation and authority

The Manager will formally declare that partial service has been achieved. The emergency organisation will be dissolved and the normal bulk water operation will resume.

1.7. Mitigation

In Table 6, various mitigation measures are given.

Table 6. Mitigation measures

Item	Expected Completion Date	Action by
Example		
Pipeline valve chambers — seismic strengthening of pipe couplings	*Define*	*Define*
Electrical panels. All secured	*Define*	*Define*

1.8. Preparedness

In Table 7, various preparedness measures are given.

Table 7. Preparedness measures

Item	Expected Completion Date	Action by
Mutual aid agreements	*Define*	*Define*
Establish communications link with clients	*Define*	*Define*
Provide sleeping quarters at emergency operation centre	*Define*	*Define*
Provision of bedding for above	*Define*	*Define*
Provision of tinned food for above	*Define*	*Define*
Emergency water supply for Headquarters	*Define*	*Define*
Emergency generator fuel supply	*Define*	*Define*
Purchase of pipeline materials	*Define*	*Define*

1.9. Response instructions for staff, including checklists
1.9.1. Pipeline maintenance section
1.9.1.1. During working hours

If practicable, the job being worked on should be made safe and preparation made for a check on the mains and branch lines, which will be organised by the Overseer.

Communications may be cut and every staff member must use their discretion as to the course of action adopted. It is desirable that staff members should report to their normal depot as early as possible. If they are working in a remote area, then they should proceed back to the normal depot following as closely as possible the

Pipeline system	*Define*
Pipeline section inspected	*Define*
Area location	*Define*
Finder drawing number	*Define*
Number of breaks	
Access availability, e.g. roads open/closed	
Inspector	
Date and time of inspection	

* Radiotelephone to emergency operations centre as soon as possible

Fig. 50. Post-disaster Pipeline Inspection Form Section A

route of the mains, noting *en route* any damage and its extent. Use should be made of the Pipeline Inspection Forms (Fig. 50) carried in vehicles.

Valves should *not* be shut unless the flow of water is a threat to life or seriously impeding rescue work. The vehicle radio should be left on, even if it is not immediately working. When radio communications are re-established, only the most urgent messages should be transmitted as briefly as possible.

1.9.1.2. Outside working hours
If possible, staff should collect food, warm clothing and a torch. It may be some time before they can return home. They should report to the normal depot and prepare for checking the mains and branches.

1.9.1.3. General procedure
* *Define inspection priorities and any operations priorities.*
* *Establish the water system status by inspecting the inlet condition of each reservoir. Then, carry out a detailed systematic inspection of the pipeline system.*

Use should be made of the Pipeline Inspection Forms Section A (Fig. 50) carried in vehicles. The number of breaks per section of line should be reported to the emergency operation centre on the radiotelephone. Section B (Fig. 51) should be filled in and returned to the emergency operation centre as soon as possible. Details of any significant damage should be noted and reported periodically by way of radiotelephone.

1.9.2. Location of pipeline materials
The pipeline materials are stored at various locations, set out below. The pipes are buried on top of the ground, with a light earth cover to stop *ultraviolet* degradation of the tape coating. The couplings and weld bands are located in the following buildings (*Define*). Care should be taken upon entry to buildings when pipeline materials are being retrieved. Each building should be examined for damage, cracks, etc.
If in doubt, do not enter.

* *Define list of available materials and locations.*

Pipeline system	*Define*
Pipeline section inspected	*Define*
Area location	*Define*
Finder drawing number	*Define*
Description of affected pipeline, valve chamber, stream crossing, etc.	....
Assessment of materials required, e.g. lengths of pipe	....
Assessment of plant required, e.g. excavators, cranes, weld sets	....
Inspector	
Date and time of inspection	

* Forward to emergency operations centre as soon as possible

Fig. 51. Post-disaster Pipeline Inspection Form Section B

1.9.3. List of welders
* *Define names and contacts.*

1.9.4. List of welding machines for hire
* *Define names and contacts.*

1.9.5. List of excavators for hire
* *Define names and contacts.*

1.9.6. List of cranes for hire
* *Define names and contacts.*

1.9.7. Excavation procedure
1.9.7.1. Get a permit!
* *Define names and contacts in city/state highway departments.*

1.9.7.2. Notify other authorities affected by the dig

* Power
* Gas
* Telecommunications
* Highway
* *Define names and contacts.*

1.9.7.3. Notify Occupational Safety and Health (if notifiable under the Government Construction Act)
* *Define names and contacts.*

1.9.7.4. Locate the underground services

* Use locators.
* Hand dig to expose services.

1.9.7.5. Check clearance of overhead power lines
If less than 4 m, the following should be contacted:

* Power authority
* Rail (if working under rail power lines)
* *Define names and contacts.*

1.9.7.6. Leave excavations safe

* Barriers and fencing
* Road signs
* Lights